The Mineral Kingdom

The Mineral Kingdom

by
Paul E. Desautels
*Supervisor, Division of Mineralogy,
Smithsonian Institution*

*with photographs
by
Lee Boltin*

*Consultant editor
for this edition
Peter Embrey
British Museum
(Natural History)*

*Paul Hamlyn
London · New York
Sydney · Toronto*

Published 1969 by

THE HAMLYN PUBLISHING GROUP LIMITED

LONDON · NEW YORK · SYDNEY · TORONTO

Hamlyn House, Feltham, Middlesex, England

SBN 600 02519 5

Printed in Italy by Officine Grafiche A. Mondadori, Verona

PRECEDING PAGES: *Copper dendrite from New South Wales, Australia.*

FOLLOWING PAGES: *Adamite from Durango, Mexico.*

Contents

Introduction

This book is meant not as a highly methodical survey of all that is known of the mineral kingdom, but as an introduction to a very broad field of knowledge through a selected sampling of certain areas—a guided tour through some of its features, rather than a textbook lecture.

It is intended for the developing interests of the amateur: the keen museum visitor hypnotized by the colour and fire of a magnificent gem, the collector who spends his holidays exploring for specimens to add to his increasingly exciting collection, the student of technology, fascinated by the odd fact that in a computerized age some of the world's greatest riches are still being brought up from the earth's interior by half-naked men using hand- and foot-power, exactly as in the time of Solomon.

Fundamentally, it is written from the mineralogist's point of view, and to the mineralogist minerals are significant not merely because they produce a trayful of diamonds at Tiffany's or the silicon wafer that makes it possible to construct a radio inside a finger ring. The mineralogist sees the mineral kingdom as a changing, developing edifice based on solid substances. He is concerned with the geological processes that formed, and are still forming, these substances, and with the highly organized internal atomic structures that result in their being, by definition, crystals. Both geology and crystallography are extremely complex sciences, but some information about them has been outlined in this book (see Chapters 2 and 7) because it is important to anyone interested in learning more about this world of solid substances.

Most of the driving force for exploring and exploiting the mineral kingdom has come from the age-old demand for the raw materials of industry and the lure of exotic and valuable gemstones. Our technological civilization is totally dependent on mineral products wrested from nature. Gems, sought to satisfy the urge for owning objects both valuable and beautiful, are closely tied to man's cultural and political development. Oceans have been crossed and new territory conquered in their quest. Accordingly, both gemstones (see Chapter 3) and industrial minerals (see Chapter 8) have been given special attention.

Even aside from his need and desire for mineral materials, man's imagination has always been stimulated by the strange and wonderful things found among them. Some attract attention because of their great beauty. Others provoke curiosity through their strange appearance or behaviour. A sampling of these striking and unusual minerals is given in Chapter 4, as well as, farther on, an introduction to classic items for collectors.

Naturally, no insight into the kingdom is complete without intimate glimpses of some of the places where noteworthy minerals have been found. The mining fields that are discussed in Chapter 5 differ in their histories and in the kinds of deposits they yield, and form a chain completely circling the globe the long way around from California to Australia.

Whatever man's contact with the mineral kingdom—for his science, his industry or his cultural pursuits—the hope is that it will have been touched on in this book to a stimulating degree.

Paul E. Desautels

Chinese bronze ornament, 5th to 2nd century B.C.

Myth, Fancy, and Fact

1

Four hundred years ago, in an age when murder by poison was a fine art, enemies of Benvenuto Cellini, the gifted Italian goldsmith, plotted to do away with him by mixing a powdered diamond with his food. Economy evidently was a factor, however, for at the last moment the assassins used a cheaper beryl instead—prompting Cellini later to record, triumphantly, that because diamond was not used '. . . the powder had no effect'.

Cellini, who dealt constantly with precious stones, should have known that in fact the powdered diamond would have had no effect either, unless it had been served up in such quantity that he would have detected it as he chewed. But in the sixteenth century man's knowledge of the properties and characteristics of minerals was obviously uncertain.

Ancient man, of course, did know of the existence of the mineral kingdom. He started very early to accumulate information about it. He learned something about working metals and setting stones. But his knowledge was so thoroughly mixed with myth, superstition and guesswork that even serious scholars of the more enlightened periods managed to hand down only a small body of reliable fact.

The most important early work on natural history was the thirty-seven volume *Historia Naturalis* written by Pliny the Elder (A.D. 23-79). In this, Pliny describes a long list of fabulous and mostly non-existent stones, some of which were supposed to have originated in the stomachs of animals, as a result of the light of the moon, or by purge from the sea. Some of his stones were male, some female. Almost all possessed some sort of occult power. They kindled fires, quieted winds, increased intelligence, and in other ways influenced the unseen forces that were believed to surround man from birth to death.

Many real stones shared these mythical attributes. An amethyst pendant suspended on a dog-hair cord around the neck was supposed to be a guaranteed antidote for snakebite. Amethysts whose purple depths showed a pale, rosy glow enabled a drinker to ward off intoxication. Diamond was considered extremely powerful; while aware of its genuine attributes of hardness and fire resistance, Pliny—a man of his time—could also credit the notion that it protected the wearer against madness and violent fear. When a diamond was to be cleaved, gemworkers of Pliny's day first steeped it in goat's blood. This was supposed to make it fragile enough to split easily.

Among other inroads into genuine science, Pliny recognized several mineral species, including one he called 'crystal', which is our quartz. With his contemporaries, he believed that this material occurred only in mountainous areas like the Alps and was formed from moisture congealed by extreme cold. Pliny's mixture of verifiable fact enriched with unrecognized fantasy was typical of the 'scientific' thinking of early periods.

The mineral kingdom abounds in curiosities that have caught man's attention from his first days on earth. One phenomenon that still excites superstitious speculation, and must have puzzled earlier men to the point of terror, is the meteorite. The ancient Jews held these mysterious objects in special reverence. They called them *Beth-el*—House of God—because they believed that the fallen fragments carried the blessings of God directly from heaven. The early Jews added other superstitions to mineral lore; among them was a story that one of King Solomon's rings was set with a stone of such peculiar power that he had only to gaze at it to know everything he needed to know at a given moment.

PRECEDING PAGES: *A polished sample of jasper— fine-grained quartz—from Oregon resembles an unreal landscape. Varicoloured iron and manganese impurities, trapped during formation, create the patterns.*

Magic rings were a fantasy common to many cultures. The Greek philosopher Apollonius of Tyana was reported to have possessed several, each set with a different metal or mineral. According to his contemporary, St. Justin Martyr, writing about A.D. 150, Apollonius could use his rings to calm the fury of the sea, hold back the winds of heaven, and tame wild animals instantly.

During the Middle Ages, the ancient superstitions formed a core around which many new ones collected. The toadstone, for example, was a widely accepted legend. It was supposed to be a stone endowed with medicinal powers, which could be extracted by those who knew the proper rites from the head of a toad. Both stone and powers were completely apocryphal, of course, yet many respected writers of the period described how the toadstone worked to soothe toothaches, cure diseased kidneys, quiet stomach disorders, and burn the skin slightly if poison was nearby.

Even as late as the fourteenth century, Albertus Magnus, reflecting current ideas in his book *De Mineralibus et Rebus Metallicis,* felt it necessary to discuss whether or not stones had spirits. His

In his famous book De Re Metallica, *sixteenth-century scholar Georg Agricola included the earliest known illustration of ore prospecting by the use of the divining rod—a method of which he disapproved.*

conclusion was that they did not, but it ran counter to the general belief.

Magical stones, meanwhile, appeared and reappeared in contemporary folklore. At the beginning of the sixteenth century an Italian named Camillus Leonardus wrote a book about gems, in which he described an astonishing stone called an alectorius. This crystal-bright stone, about the size of a large bean, was allegedly carried in the body of a cock, and was mature and ready to be extracted when the cock refused to drink. Milo of Croton, a strong man of the sixth century, was supposed to have carried one constantly and to have lost his strength only when he lost the stone.

The alectorius had numberless virtues. It banished thirst, made a wife more appealing to her husband, and gave eloquence and persuasive power to its owner. It protected domestic peace and harmony, and attracted honour and victory. Clearly it would have been a most desirable possession—had it existed.

Even during the Renaissance, that period of 'new learning', men like Paracelsus continued to expound a mixture of fact and fiction about the mineral kingdom. Paracelsus was a Swiss physician,

alchemist, chemist and lecturer, who introduced into Europe various new and effective mineral medicines—mercury, iron, and arsenic among them. However, he also believed in a metaphysical relationship between the mineral world and the planets and neighbouring stars. According to these theories, the metals and minerals on earth had been thrown off by other planets, while the crystalline gem minerals, such as emerald and rock crystal, came to earth after having condensed on the nearer stars.

We still preserve traces of the old metaphysical notions that tie gems to the motions of the planets and stars, and suggest that they have control over human events. People still wear their birthstones, though not many moderns would admit that by so doing they are offering a hostage to fate. The Hope diamond, which now flashes gloriously in its display case in the Smithsonian Institution in Washington, was for years romanticized in the press as a jewel in whose wake misfortune trailed.

Yet despite the embroidery of fact with legend and folklore, the ancients managed to contribute much more than superstition to the development of mineralogy and metallurgy. Early experimentation produced a great deal of useful information. There is evidence that many early civilizations were familiar with the working of metal ores.

The shift of the centre of civilization westward into Europe, some time before 1400 B.C., was most likely stimulated by the availability of iron ores and forests for the developing Iron Age. The Incas were working copper about 1000 B.C. King Solomon derived some of his glory from successful copper-mining operations carried on about 950 B.C. in what is now a desolate Israeli desert. Ancient China was certainly preparing metals for decoration and practical use at the same time as the Egyptians. The Greeks and Romans made

FACING: *Crystal blades of blue kyanite from St. Gotthard, Switzerland, formed in a metamorphic environment.* ABOVE: *Botryoidal mimetite from a sedimentary environment in San Luis Potosí, Mexico.*

a

b

c

d

wide use of copper and bronze for tools, coinage, weapons and ornaments. In the New World, when the Muzo Indians of Colombia were conquered by Gonzalo Jiménez de Quesada in 1537, they offered him emeralds as a gift. These Indians, and their ancestors before them, had worked their emerald mines for so long that the history of the operation was lost in the mists of time. These isolated illustrations show that man began very early in his history to develop knowledge of the mineral kingdom and to put it to practical use.

The long-accumulated, confused and interwoven mass of fact and fantasy about minerals, gems, mining and metals prevented the ancients from organizing any logical scheme of their nature. The facts, nevertheless, were there for later investigation when the political and cultural atmosphere was ready for development of scientific thought.

WHAT IS A MINERAL? A mineral is, first of all, something not belonging to the animal or vegetable kingdoms. No definition of a mineral has yet been proposed that is all-inclusive and satisfying to everyone. With various slight modifications and elaborations, the traditional definition, appearing in most texts, states that a mineral is 'a chemical element or compound occurring in nature as a result of inorganic processes'. This is serviceable only if it is not interpreted narrowly. To gain a real understanding of it, the statement needs explanation and expansion.

The existence of chemical elements is one of the basic truths of all modern science. We know that the material universe is made up of a few more than a hundred fundamental elements. All the different kinds of things in all the kingdoms— animal, vegetable, mineral—are composed of these elements. They may be made of the pure element or of very complex combinations of elements.

Some elements, such as gold, copper, iron, carbon and oxygen, are well known. Others, such as lanthanum, xenon and polonium, are primarily laboratory curiosities.

When chemical elements combine with each other, their atoms combine in definite proportions. An atom can be imagined as the smallest possible particle of an element. Atoms of each element have a characteristic combining ability; this controls the proportions in which atoms combine to form larger groupings such as crystals or molecules.

For example, the element sulphur exists by itself in certain mineral deposits in the crust of the earth. It is a bright yellow, soft, resinous-looking substance that burns easily with a blue flame. However, sulphur can exist in combination with other elements, such as antimony. Usually three sulphur atoms unite with two antimony atoms. The chemist indicates this with the formula antimony$_2$ sulphur$_3$ or, using his customary symbols, Sb_2S_3. The mineral name of this is stibnite; it is a bright, silvery, soft, metallic compound.

Sulphur may also combine with iron, thus

FACING: *Medieval medicine from* Hortus Sanitatis *(1483)—(a) lapidary prepares stones; (b) toadstone is extracted and (c) it draws out poison; (d) bloodstone cures nose-bleed.* ABOVE: *Alchemist's triangle symbolized spirit, soul, body.*

17

forming the mineral pyrite, FeS_2. If the same atoms are hooked together in a different arrangement the mineral marcasite, also FeS_2, may be formed. Because they have the same composition although their atoms are joined in different ways, pyrite and marcasite are called polymorphs. There are other cases in which different element combinations—for example, $CaCO_3$ (calcium carbonate, called calcite) and $MnCO_3$ (manganese carbonate, called rhodochrosite)—have their atoms joined in identical patterns. Such minerals are called isomorphs.

The first part of the traditional definition, then, speaks of elements and compounds, implying that the way the atoms combine to form different patterns is of primary importance. This idea is the basis of all really useful systems of grouping or classification of minerals. In all branches of science, attempts constantly are made to classify available facts. Unless they are related and compared with each other, however, such facts tend to remain isolated and sterile. Related to others, they lead to new ideas, new laws and new understandings.

Attempts at trying to order and describe natural geological materials began early. About 300 B.C., Theophrastus, a Greek philosopher, wrote his book *On Stones*. Pliny's books, almost four hundred years later, included passages on gemstones, pigments and important ores. In 1556 the famous *De Re Metallica* of Georg Agricola (the pen name of Georg Bauer), German scholar and mining authority, was published. His book brought up to date man's knowledge of mineralogy, mining and the uses of metal ores.

The science of crystallography was given a bold start by Nicolaus Steno in 1669 with his astute observations on crystal measurement. By the middle of the eighteenth century, the allied sciences of mineralogy and chemistry were under intensive investigation all over Europe. The transition was being made from superstition, myth and conjecture about minerals. Increasingly, critical observation would now be applied to the mineral kingdom.

In 1774, the renowned Abraham Gottlob Werner, professor of mineralogy at Freiberg, Saxony, published his essay on the *External Characters of Minerals*. His ideas about the separation and grouping of known minerals by their external characteristics were spread rapidly by students, who came from all over Europe to study with him. The system of separating and arranging minerals according to their colour, hardness, inflammability and so on, seemed very logical. Relationships among the

minerals were easier to see and new predictions about their chemical and physical behaviour were easier to make.

In this same period, the French mineralogist René Just Haüy established an approach to the study of crystals that led directly to our modern understanding of them. His basic tenets are discussed in Chapter 2.

Finally, by the beginning of the nineteenth century, the Swedish chemist Berzelius had outlined a chemical classification of minerals. It was he who developed a theory of positive and negative kinds of atoms, which attract each other to form compounds. He showed that minerals are such compounds. Berzelius went on to publish a system of mineral arrangements based on many chemical

FACING: *Ancient goldwork in a figurehead from a Viking ship*. THIS PAGE: *Stalactitic masses in widely differing forms*—TOP, *Marcasite crystals from Montreal, Wisconsin;* ABOVE, *Green malachite from the Congo*.

19

analyses of minerals. In it, all the minerals of similar chemical composition, such as mercury sulphide (HgS) and lead sulphide (PbS), appeared as near neighbours.

It was not until 1854 that James Dwight Dana, professor of mineralogy at Yale, set the chemical classification of minerals on a firm footing. In 1837, Dana had published his first *System of Mineralogy*, using 'a system of nomenclature constructed on the plan so advantageously pursued in Botany and Zoology'. However, when the time came for the fourth edition of the monumental work in 1854, the author was forced to abandon this biological approach, which had been first established for botanical classification by Carolus Linnaeus of Sweden in the eighteenth century. More was known about the chemical nature of minerals; to show their relationships properly, Dana classified them in a purely chemical system. To this day, the primary classification system of the mineral species is based on their chemistry. From the 1854 edition of the *System* down through the current seventh edition, the increase in information about mineral structures obtained through use of X-rays has brought about the only major modification in Dana's system.

The discovery of X-rays in 1895 was to lead, in a

few years, to development of the most powerful tool available for mineralogical studies. In 1912 Max von Laue, who had been a student of Wilhelm Konrad Röntgen, discoverer of X-rays, began experimenting with the application of X-rays to mineral crystals. He proved that X-rays were small enough to detect spaces between the atoms of which a crystal is made, and could be used to measure the arrangement of these atoms.

With the development of the science of X-ray crystallography and the investigation of mineral crystal structures with X-ray gadgetry, it became obvious that chemical composition alone was not

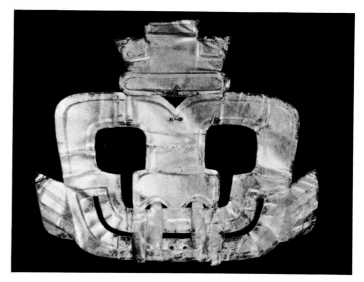

the answer to the nature of mineral species. The arrangement of their atoms was revealed and recognized as being of extreme importance. Thus, pyrite and marcasite, as before noted, are polymorphs—alike in chemical composition. But the different arrangement of their atoms results in their very different physical appearance and behaviour.

Systems of mineralogy in wide use, prepared by prominent modern mineralogists, include those of Max Hey of England, Hugo Strunz of Germany, and Clifford Frondel of the United States. All of them take into account the composition, and some the atomic structure, of mineral species. The systems are in many respects similar, differing mainly in incidentals.

In any classification system there is always a problem in applying names to the species. Minerals unfortunately have been overburdened with too

many meaningless names accumulated through the centuries. For example, under the classification system devised by Sir John Hill in London in 1748, the mineral species pyrite was given several different names, based only on differences in shapes of the crystals—an effort at clarification that merely encouraged confusion.

The modern method of chemical-structural classification is far simpler and much more precise, although the names of many minerals in it are still scientifically meaningless. The naming of a new species has always been the privilege of the scientist first describing it. Consequently, species are named somewhat haphazardly for superficial appearance, noteworthy characteristics, mode of occurrence, for people, places, and even political effect. Olivenite, for example, is named for its colour, though a name based on its chemical-structural analysis would be

far more significant. Reedmergnerite was named for two men, Reed and Mergner, of the United States Geological Survey. In recent years the species pravdaite, subsequently discredited, was named for the official Russian newspaper *Pravda*.

Nevertheless, to be given any name, and to be accepted as a distinct and separate species, a mineral must prove to be chemically or structurally different from any other known mineral. The distinguished mineralogists who make up the International Mineralogical Association have established a Mineral Names Commission and set high standards for the scientific proof of a new species. In addition,

research is carried on in many places to determine the status of those species that are in doubt. In older editions of Dana's *System of Mineralogy* more than three thousand mineral species are listed. The accepted number today is much nearer fifteen hundred. Several new species are discovered and approved each year, but at the same time several others are likely to be demoted to the status of mere varieties.

MINERAL OCCURRENCE: As noted earlier, the commonly used definition for a mineral requires that it should occur in nature. This seems an obvious necessity until we remember that minerals are,

after all, chemical elements and compounds. Many research scientists are engaged these days in preparing various kinds of chemical substances in the laboratory. Many of these substances are counterparts of some found in nature. It is common practice, in fact, for a mineralogist to prepare in a laboratory the species that he is studying in nature. This may be done to determine the natural temperature, pressure or other conditions under which the mineral originally formed. It may even be necessary to manufacture a mineral because the natural supply is too scarce to provide enough specimens for a complete study. Prepared as a synthetic laboratory

Sedimentation creates rich patterns. TOP: *Mud and limestone sediments form Italian marble slab.* ABOVE: *Sediments form flat marcasite concretion from Mazon Creek, Illinois—noted for fern fossils.*

stand-in, the mineral can be studied by proxy.

Industry, of course, is always interested in the manufacture of synthetic minerals. Manufacture often creates a purer, more economical product than can be obtained by mining and processing the ores. Mineralogists have been devising techniques for producing synthetic minerals in the laboratory for more than two hundred years. These processes have been watched closely for possible use in making synthetic gemstones, since gems produce a high financial return for their weight and are difficult to find in nature.

Several problems arise if we hold strictly to the phrase 'occurring in nature'. The last mineral species listed in the second volume of Dana's *System of Mineralogy* is calclacite. This white, hair-like mineral was first found as a coating on limestone rocks and fossils stored in oak museum cases. Apparently, acetic acid derived from the oak wood of the cases, acting on these stored rocks over a long period of time, had produced the new species. Although the process was natural enough, it would not have happened without the intervention of man, who accidentally brought together the raw materials. Perhaps the mineral species butlerite has more claim to natural origin but it too, and several others associated with it, are subject to question. Butlerite is a deep-orange, glassy mineral occurring in small crystals. It is found at the United Verde copper mine at Jerome, Arizona. In this deposit it resulted from the gases formed during the burning of pyrite-bearing ore bodies. The process is a natural one, but it is likely that the underground fire would not have started had it not been for the activities of man.

When discussing the origin of solid parts of the earth's crust, the terms 'rock' and 'mineral' frequently become confused. For most purposes it

is convenient to think of a rock as a solid made up entirely of a mixture of small bits of two or more different minerals. Sometimes a very large deposit of a single mineral species is accepted by the geologist as rock. In any case, these are solid substances that have occurred because the surface of the earth has been in a constant state of turmoil for untold centuries.

By human standards of time, of course, all the changes occur in very slow motion. Mountains

have been thrust up, eroded away, cast into the sea. The process has repeated itself around the globe again and again and again. Parts of the crust have been repeatedly raised, lowered, buried, heated, cooled, flooded, ground up, compressed, carried here and there and never left in peace. As a result, it is a relatively rare thing to find a chunk of pure mineral that has not been intimately processed into some mixture of mineral grains which we call rock. Some of these natural processes, however, tend to isolate minerals and channel them into concentrated deposits. When this happens they can be mined, and become valuable to man.

WHERE MINERALS AND ROCKS FORM: Assuming that we ignore synthetic minerals and those of questionable origin, there remain perhaps eighteen hundred others that are found in nature. Geologists tend to group the environments in which minerals form as magmatic, sedimentary and metamorphic. Meteorites have a non-earthly origin and represent a special environment.

Minerals and rocks found in a *magmatic environment*

are formed from melted, cooled material and are called igneous rocks. Magma, the fiery liquid originating deep in the earth, now and then surges toward the surface. It almost seems that inconceivable internal pressures drive the molten liquid upward. Usually the magma stops short of the surface and cools, more or less slowly, leaving deposits that, because they formed within the earth, are called intrusive rocks. Sometimes the magma succeeds in reaching the surface and bursts forth into the air with a magnificent display of fireworks, devastating all about it with a roaring torrent of ash, fire, smoke and lava.

One fiery upsurge occurred on 20 February 1943, near Michoacan in west central Mexico, where a farmer ploughing his field was interrupted by a visit from the devil—or so he thought. From a fissure in the field smoke began to rise, and a roaring sound filled the countryside. The farmer was, in fact, witnessing the birth of Paricutin, a typical volcano. The volcano eventually spewed out enough ash and lava to devastate many square miles, destroy the local villages and build a towering cone over one thousand feet tall.

Rocks that have formed on the earth's surface from cooling lava are called extrusive rocks. Lava thrown out by volcanoes cools so rapidly that separation of individual mineral grains has little time to occur. Lava rock, then, is not grainy but smooth-textured, sometimes even glassy. If glassy, it is called obsidian, and if dark in colour but fine-grained, it is called basalt. One of the best examples of an occurrence of basalt is the Columbia Plateau in the northwestern United States. It was created by a long series of lava floods, which piled basalt layers more than five thousand feet thick over two hundred thousand square miles of Oregon, Washington and Idaho.

FACING: *Typical striations on jumble of pyrite cubes from Leadville, Colorado.* ABOVE: *James D. Dana. Before his* System of Mineralogy *simplified classification, pyrite—like other minerals—had many names and confused identification.*

25

Minerals and rocks found in a *sedimentary environment* are formed from those gravels, sands and clays deposited—most frequently by water, but sometimes by wind or by glacial ice—in slowly building layers. Actually, the great bulk of rocks of the earth's crust is magmatic, formed from the cooling and solidification of molten material. This, however, is not obvious from a cursory examination of the rocks that show through the soil and plant covering that dusts the earth's surface. Almost all the rock formed from molten material is covered by a thin skim of sedimentary rock. This develops because surface rocks are constantly under the attack of weathering and erosion. Some products of this erosional attack—mud, sand, gravel—are carried off and deposited near and far after forceful separation from the mother rock. The rate of erosion is sufficient to lower the exposed solid parts of the earth one foot every thirty thousand years. A blanket of these ever-available sediments is always being deposited somewhere.

Ultimately the sediments are converted to rock by compression resulting from the weight of other sediments on top of the pile. This conversion, or lithification, can also be caused by cementing of the sediment particles. Cementing agents, usually quartz, limonite or limestone, are carried in by water solution and deposited between the grains. Erosion may carry away some of its products in solution, and these solutions evaporate. Large deposits of limestone, gypsum and other minerals trace their origin to such sedimentary processes.

Minerals and rocks found in a *metamorphic environment* have formed by changes in already existing rock. When any kind of rock is exposed to heat, pressure, liquids and gases, it is changed—it becomes a metamorphic rock. Between the temperatures and pressures of an infernal molten

environment and the watery grave of a sedimentary environment lies a great range of changing combinations of conditions. Changes in the existing rock to meet ever-changing environmental conditions create new kinds of metamorphic rocks and subsequent new minerals in the composition of these rocks.

The only solid-object samples we have that do not come from our planet are *meteorites*, which shoot into it from space. Meteorites are special kinds of rocks formed under special conditions. However, minerals of which they are made are the same kinds as those occurring on the earth. These fascinating objects can be conveniently divided into three groups: siderites (metallic), siderolites (stony and metallic), and aerolites (stony).

FACING: *Devil's Post Pile, California. Slow cooling of thick lava layer produced these accidentally geometric basalt columns, which are not crystals.* THIS PAGE, TOP: *Microscopic butlerite crystals.* BELOW: *Gypsum from Chihuahua. Both are true crystals.*

By applying the same techniques used for the study of earthly rocks and minerals, the compositions of meteorites and the conditions under which they were formed can be determined. This information is valuable at a time when so much attention is being turned to space and planets. Diligent study may reveal geological processes which do, or do not, differ from our own; we may learn whether these processes occurred on larger planets or smaller, using raw materials similar to or different from ours.

HOW MINERALS AND ROCKS FORM: Whether the origin of rock be magmatic, sedimentary or metamorphic, there are certain basic processes by which the rocks and the minerals that make them up are formed. The most important of these are: crystallization from molten material; sublimation, or crystallization directly from a gas to a solid; formation from water solutions; metamorphism and metasomatism; weathering.

The kinds of minerals and rock mixtures that form depend in large part on the conditions under which they form. Because of this, there is an all-important law governing the entire event: At the time of its formation a mineral must be in equilibrium with its environment. This means that if pressures or other conditions change, even momentarily, then the kinds of minerals forming may also change. Even if the minerals have already formed before the change in growing conditions, they will still change in composition or structure. To exist in the end, the mineral must be perfectly adjusted to, or at least in satisfactory agreement with, its surroundings.

Crystallization from molten material: The kinds of materials formed from molten rock (or by any process for that matter) depend on the elements available. Surprisingly, of the approximately one hundred known chemical elements, eight constitute 99 per cent of the earth's rocks. Just two of these—silicon and oxygen—make up 74 per cent, leaving only 25 per cent for the other six—aluminium, sodium, calcium, iron, magnesium and potassium. This list easily accounts for our commonest commercial metals, aluminium and iron. Copper, zinc, gold, nickel, chromium and others are comparatively rare; in fact it is remarkable that nature has concentrated, in individual deposits, sufficient quantities of them to make it feasible for us to mine and use them.

As a magma cools and begins to form the solid crystals of minerals that make up igneous rocks, the liquid part remaining becomes richer and richer in less common elements and in water, carbon dioxide, other gases and certain acids. This residual hot liquid eventually cools and crystallizes in a unique manner. These last-formed minerals tend to be very coarse crystals, many of them of unusual size and composition. Such deposits are called pegmatites, and may range in size up to bodies that are thousands of feet long and hundreds of feet thick. Feldspar and mica are mined commercially from pegmatites, as are tourmaline, beryl, quartz, topaz and other gem crystals. Pegmatites

are also the source of many beautifully crystallized mineral specimens that become collectors' items. Over one hundred different minerals have been found in pegmatites.

Sublimation: Some of the extremely hot gases in magma are released if they can find an avenue of escape to the surface through a gas vent, or fumarole. Some remain as gases and pass into the atmosphere. But frequently the contents of these gases, at the cool surface, pass directly from the very hot gas state to the solid state, without cooling through a liquid stage. This process is known as sublimation.

Sulphur (S), fluorite (CaF_2), realgar (AsS), halite ($NaCl$), hematite (Fe_2O_3) and others have been found formed directly from hot gases by sublimation. Commercial ores are seldom deposited this way, but sublimed sulphur and halite have been mined.

Formation from water solutions: Minerals that form from water solutions are known as evaporites. In every case evaporite deposits occur when mineral-bearing water evaporates. In arid and semi-arid regions, lakes, or large bodies of water left by drainage into a basin, are very temporary. They evaporate and leave behind a mineral crust. Water

FACING: *Specimen of obsidian—glassy volcanic rock— typical of those from Mexican occurrences.* ABOVE: *A night-time eruption of Mexico's Paricutin. Volcanic bombs may contain rocks from great depths, providing valuable new study material.*

accumulates again with the rains, and the annually repeated process gradually builds up thick deposits of evaporite.

The great Atacama Desert of Chile is an extremely arid area, six hundred miles long, on the border of northern Chile and Peru. In this desert, one thirty-second of an inch of rain for the year is considered above average. In some places it never rains. Here there are very rich deposits of soluble copper minerals and also very large beds of soluble nitrates. At one time this area was the world's major source of nitrates for fertilizer and for the manufacture of explosives.

With normal rainfall, such deposits would long since have been dissolved and run off into the Pacific Ocean. As a matter of fact, the peculiar evaporation process by which they are thought to form could not operate. In arid regions like the Atacama Desert, underground waters carrying dissolved minerals seep slowly upward toward the surface and quickly evaporate, leaving the minerals deposited on the surface. Constant upward seepage and continual rapid evaporation gradually build the surface deposit, and no rain washes it away.

Underground water, with its dissolved mineral burden, can also migrate into underground caves or other openings. As the liquids drip into these cavities the changes in temperature and pressure cause the water to evaporate or its contents to change by giving off gases. The resulting deposits form stalactites, stalagmites, pillars and other subterranean sculptures.

Metamorphism and metasomatism: According to the previously stated law, a mineral must be in equilibrium with its environment. When nature, in its laboratory, changes some rocks into others, it operates in one of two ways: new materials are added to the original rocks, or the original materials merely are rearranged. Metamorphism occurs in both ways.

Geologists differentiate between contact metamorphism and regional metamorphism. The first occurs when the earth strains and shifts, and new intrusions of magma force up into already stable country rock. Heat, pressure and the dissolving action of actual contact with magma cause the surrounding rock to change. Also, hot liquids and gases emanate from the magma and penetrate the surrounding rock. This causes additional changes, which are more extreme near the magma and taper off with distance from it. Metamorphism caused by penetration of these hot liquids and gases is given the special name of metasomatism.

Regional metamorphism describes itself. It is caused by large-scale changes in the earth's crust, and may extend over thousands of square miles. The buckling and thrusting of the crust that resulted in the dramatic birth of the Appalachian Mountains millions of years ago shortened the distance between what are now Baltimore and Pittsburgh by perhaps two hundred and fifty miles. The forces necessary to move such vast tonnages of rock and thrust them thousands of feet into the air are incomprehensible. Yet this occurs again and again. Little wonder that the rocks in such regions are stressed beyond endurance and change themselves to meet the demands made upon them.

Dynamic motion of the earth's crust is not a requirement for regional metamorphism. The slow, undramatic, but continuous building of layers of sediment, century after century, can have the same effect. Water pressure five miles under the sea is about twelve thousand pounds per square inch. Imagine the pressure under five miles of rock sediment, which is three times as dense. These rocks will alter to suit the new environment.

a

b

Each of the three geologic environments in which rocks form produces many different kinds of rocks. Shown here, from a metamorphic environment, are: (a) marble and gneiss from Moravia; (b) schist, granulite, meta-diorite

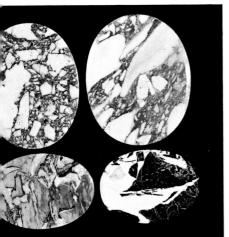

c

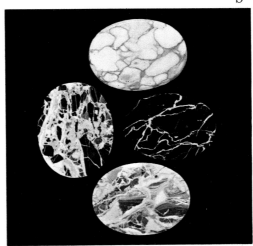

d

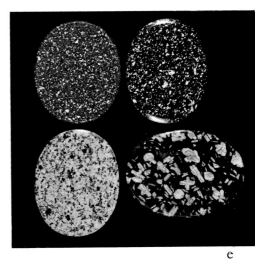

e

from Egypt and Moravia; (c) brecciated marble and limestone from Italy and Spain; (d) brecciated marble and limestone from Italy, Sicily. From a magmatic environment: (e) porphyry and granite from Italy, Egypt, Rumania. From a sedimentary environment: (f) limestone from Sicily and Austria; conglomerate from Italy; (g) limestone from Spain, Sicily, Italy, Bohemia.

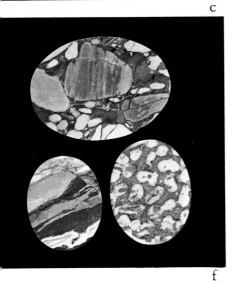

f

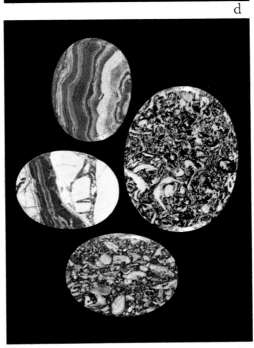

g

31

Weathering: There is a tendency to accept the notion that rainwater 'just wears away the rock'. By itself, water works very slowly in its attack on rock—unless, of course, it rushes past carrying sand grains and gravels that grind and scratch as they tumble along. The devastating effect of rain as a weathering agent is due to the dissolved gases—oxygen and carbon dioxide—which it carries along from its falling passage through the atmosphere. These dissolved gases turn rain into a very weak acid which, over a period of time, chemically attacks the various minerals, converting them to new chemical substances which in turn attack their neighbours. Thus the original integrity of the rock, ore or mineral is destroyed.

The rich copper deposits of Arizona became commercially valuable only because of centuries of downward chemical leaching, or dissolving, of low-grade copper minerals. These were gradually redeposited in more concentrated form deep in the deposit. Later, as erosion progressed, the deep, enriched ores were exposed for exploitation of the scarce and valuable metal.

All of the geological processes and environments mentioned are related to each other in unbridled complexity. Feldspar is one of the most abundant minerals formed from magma. When it is exposed and weathered it is readily converted into clay. Deposits of sedimentary clays, when they are deeply buried by other layers of sediments, react to the great increase of pressure and temperature by changing to metamorphic muscovite and garnet. Then they are stable in their new environment until erosion starts the chain again.

It is obvious now that expansion of the definition 'occurring in nature' as a result of 'inorganic process' is a science in itself.

The mineral kingdom is extensive, varied and still largely unknown. Science is exploring it actively, industry is exploiting it eagerly—but the chances are that nature, which is constantly busy changing and renewing it, will always be one step ahead in this chain of events. We have still so much to learn about this planet, and now mineralogists will be receiving samples from the moon, and eventually from the planets beyond.

ABOVE: *In Bryce Canyon, Utah, sculptured appearance of eroded remnants was produced by uneven weathering.* TOP: *A geode lined with amethyst crystals from Rio Grande do Sul, Brazil.*

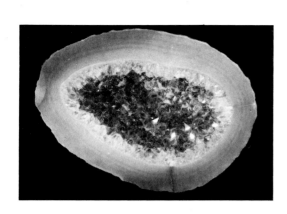

Flowers of the Kingdom

2

For the physicist the words 'crystal' and 'solid' carry the same meaning, because all solids have the internal organization of atoms comprising crystal structures. Since all minerals are solids and all solids are crystals, then all minerals are, obviously, crystals. The study of crystals therefore is essential to an understanding of all the inhabitants of the mineral kingdom.

The term 'flowers of the kingdom' was first used by the renowned crystallographer Abbé René Just Haüy of France. He was referring to externally well-formed natural crystals of minerals, those remarkable objects that often seem as brilliant and sparkling as the gemstones that may be cut from them. It is not unusual for natural crystals to be so geometric in appearance that they seem to have been cut by some master craftsman for use as models in a study of solid geometry. The formation of crystals is an expression of natural laws, and outward evidence of marvellous atomic activity that has created their intricate structures.

The word crystal is derived from the Greek word *krystallos,* which in turn comes from *kryos,* meaning icy cold. The Greeks believed that crystals of quartz resulted from ice having frozen so hard it would never thaw out. Although the name was first used only for quartz, it came through the centuries to include all crystals and even some kinds of manufactured glass. This glass, used for goblets and other tableware, looks like quartz, but it is not crystal at all. A crystal is a solid, geometric object which has, under suitable growth conditions, an external form composed of a number of related flat faces. Its form is the outside expression or evidence of a very highly organized internal arrangement of atoms in tiny units stacked together. All true crystal is made this way.

All so-called crystal, then, is not crystal. The word has a very precise meaning, which is universally understood in the various sciences. Why is it incorrect to call cut-glass tableware crystal? It certainly seems to be solid, and it has an appearance similar to that of some natural crystals. It might actually be truly crystalline if it were a solid substance. But, strange as it seems, glass is a liquid.

The physicist recognizes three forms in which the materials of the universe can exist: solids, liquids and gases. The major differences in the three states are due to the differences in energy of their atoms. Gas atoms are the most energetic, active and widely spaced, and have little control over their mutual destinies. Liquid atoms, on the other hand, try to group themselves but, being still too energetic, slip past each other. The atoms are spaced somewhat apart and lack overall organization among themselves. Solid atoms have comparatively little energy; they are closely spaced, and therefore they force each other into organized patterns.

Glass is a liquid, then, which becomes glass when the atoms in the liquid are suddenly reduced to the energy level of a solid before they have a chance to get organized. Atoms in glass become locked in a chaotic mass. In time they may organize themselves and become true crystal solids. But this happens far too slowly to be noticeable.

The real beginning of the science of crystallography came in the sixteenth century with the publications of the mathematician, Johannes Kepler. He wrote dissertations on the subject of geometrical solids and derived mathematically a series of solid forms identical with those found among minerals. In 1611 he published a paper that explained the uniformly hexagonal shape of snow crystals as an orderly arrangement of small units composing the crystals. About the same time, Athanasius Kircher

PRECEDING PAGES: *Crystal models, used in the study of natural crystals, are cut from many materials. These models are of the high-quality glass popularly, but incorrectly, called 'crystal'.*

advanced the proposition that there was some kind of internal force in crystals causing them to develop outward along certain fixed directions.

In 1669 Nicolaus Steno established the first law of the science of crystallography. This fundamental law—The Constancy of Interfacial Angles—seemed simple, but was actually a very profound observation. It stated that for any crystals of a given mineral, the angles measured between the same sets of crystal faces were always exactly the same. Any crystal of quartz, for instance, no matter how lopsidedly grown or malformed or large or small, when measured with a simple instrument called a goniometer—*gonio* meaning angle and *meter* meaning measure—could be proved to have universally identical sets of angles. Steno's work was based on an accumulation of rather crude measurements on quartz crystals. But by 1783 Romé de Lisle, a French crystallographer, had measured an enormous number of crystals with some accuracy and substantiated Steno's law.

The science of crystallography received its final push up to modern times from the Abbé Haüy during this same period. Haüy, honorary canon of the Cathedral of Notre Dame, was a professor at the Museum of Natural History in Paris. In his *Traité de Minéralogie* he was the first to develop the important fundamental notion that every crystal is built up by an orderly stacking of tiny sub-units, and that the final shape of the crystal depends both on the shape of these small units and the way they are packed together.

Startling scientific discoveries are rarely made by accident. They are made rather by men who are mentally prepared to observe and interpret anything they experience. However, legend has it that Haüy, while examining a collection at the home of a friend, dropped a specimen of calcite.

TOP: *Pyrite cubes—an intergrown mass from Ambasaguas, Spain.* ABOVE: *Crystals of amazonite, a feldspar variety, from Pike's Peak, Colorado.*

that at the surface of the crystal the units could be progressively staggered to form a developing series of steps in the direction of the crystal faces. How, then, could the faces be so smooth if they were composed of steps? Since the units were very small the eye could not distinguish the steps. As a matter of fact, later X-ray work showed the units to be about one hundred millionth of an inch long; the unaided human eye can normally distinguish only objects about one million times larger.

Out of this theory, Haüy next developed a system of axes, which can be used to measure a crystal face by the way each face cuts each axis used for describing each particular crystal. He developed this system into the Law of Simple Rational Intercepts, which has been known ever since as Haüy's Law.

In 1848 Auguste Bravais, a mathematician, showed the possible patterns that could be made from a series of points arranged symmetrically in space. These fourteen possible groupings, known as the Bravais lattices, are still essential for understanding modern X-ray crystallography. In 1912 Max von Laue used X-rays to study crystals and was able to confirm that everything Haüy had proposed was essentially true. Crystals are actually made up of very tiny units in an orderly arrangement. The units are much smaller than Haüy ever dreamed, but they follow his ideas closely and make his law valid. The arrangements closely parallel the Bravais lattices. Later work by the father-and-son team of English physicists, W. H. Bragg and W. L. Bragg, refined von Laue's work; they showed that these crystal lattices existed and were caused by precise, orderly placement of the atoms of a crystal.

Gathering together the ruins, Haüy noticed that the broken pieces were all quite similar and resembled other calcite crystals he had seen. With an exclamation of '*Tout est trouvé!*'—All is discovered!—he returned to his laboratory and systematically demolished all the different shapes of calcite crystals he could find in a successful effort to prove his point.

No matter what the shape of the parent crystal, the broken-down pieces all had faces at the same angles, each one being a rhombohedron—a parallelepiped whose faces are all parallelograms. Haüy found that he could continue cleaving each of these pieces until it was too small to handle, with the same results. He reasoned that eventually there would be a limit to the size of the pieces that would break out in this regular shape, and concluded that every calcite crystal is made of thousands of very tiny rhombohedra.

Not stopping there, Abbé Haüy boldly explained the fact that not all calcite crystals are the same shape on the outside. He did this by assuming

Chemical elements are not uniformly distributed throughout the crust of the earth but have been concentrated by various geological processes in

particular places. These concentrations or deposits constitute the natural resources of mineral materials, which we exploit for science and industry. It is in these concentrations that a sufficient number of atoms of elements accumulate to form crystals. The crystal-growing process begins and the atoms go through a series of energy and position changes that will ultimately produce the solid. Let us see what these processes are.

Much of the activity, physical and chemical, that we observe on the earth requires a loss of energy. An automobile runs by using up or 'losing' energy stored in its petrol tank. An alarm clock runs and rings by using up the energy stored in its wound springs. In the same way, a crystal can form by the accumulation of atoms only if

the atoms lose some of their energy. Melts, solutions and gases are the source of all the very active atoms that make up crystals. Atoms found in hot, melted materials as gases or in solutions are extremely active, moving about at tremendous velocities. They are packed with energy, which exhibits itself as rapid motion. The atoms that are locked up in solids are not available unless the solids can be melted, dissolved or vaporized, or unless energy can in some way be added to the atoms to turn them loose.

Crystals are formed when cooling gas or vapour atoms slow down, get closer together, grasp each other with strongly attractive forces, and become locked in a regimented order. Every snowfall illustrates this beautifully. When the air,

FACING: *Abbé René Just Haüy was the first mineralogist to suggest that crystals had orderly internal structure.*
ABOVE, LEFT: *Contact goniometer, 1780.*
RIGHT: *Accurate reflecting goniometer, c. 1900.*

39

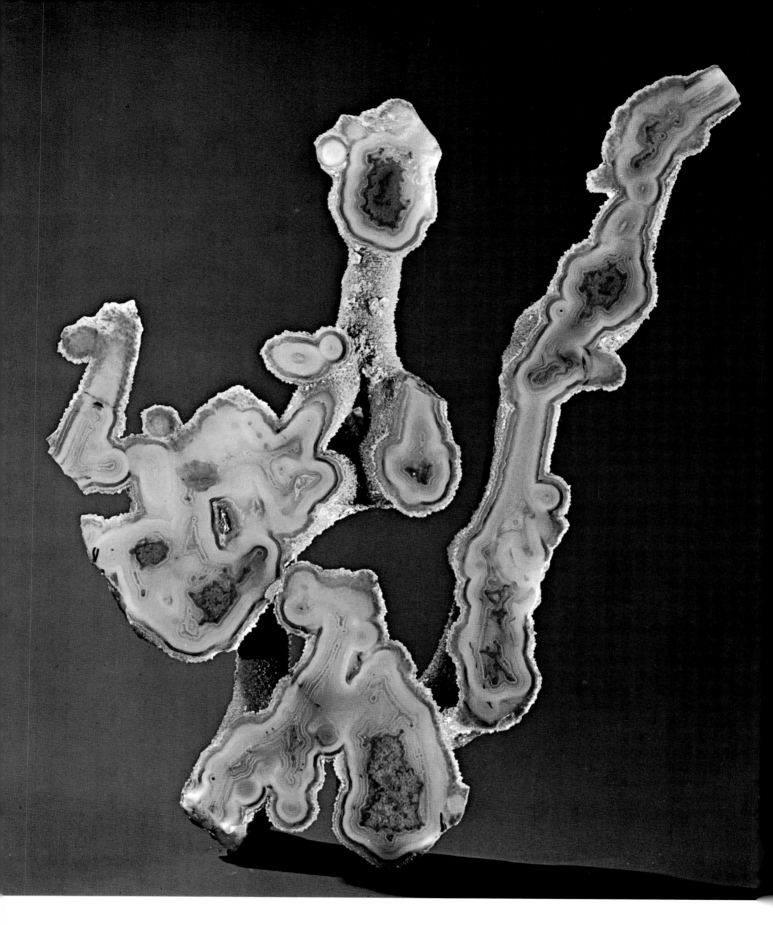

heavy with water vapour, cools sufficiently, atomic groups composing the water vapour slow down and finally lock each other into the beautiful hexagonal patterns of snow crystals. Sometimes the shift from vapour to crystal occurs as a two-step event. First the atoms slow down, so that the vapour becomes a liquid. Then the atoms continue to decrease their activity until they actually grasp each other tightly in a solid embrace. The process of losing energy and organizing a solid structure is the same for atoms in melts and solutions.

Atoms in solution, like those in gases, can be formed into crystals by cooling the solution to remove energy. Crystals can also start to form when some of the water, or other solvent, is removed by evaporation. The atoms of water in a water solution of salt, for example, have an average energy of motion. However, a few of them are more energetic than the average. Now and then some of the over-energetic ones move so rapidly that they escape the solution and move into the surrounding air, taking along their extra supply of energy. This leaves the entire solution a little poorer in energy than it was. The smaller, less energetic amount of water still left in the solution interferes less with the process of forming salt crystals. The atoms composing the salt move closer together and formation of solid salt crystals begins.

The earth is constantly surrounded by an envelope of gases and is constantly evaporating water from its surface. The earth is also constantly forming new solutions by movement of water on and in its crust, and is periodically forcing molten rock towards the surface. Thus there is always a ready supply of vapours, solutions and melts for the continuous formation of new crystals.

Because it is possible to reproduce some of nature's conditions in the laboratory, a number

of rather ingenious methods have been invented for growing synthetic crystals. The only differences seem to be that nature uses less pure materials, operates on a vastly larger scale and has unlimited amounts of time.

In the laboratory, good crystals are usually grown from melts or solutions. One of the basic techniques used is the flame-fusion process invented by a French chemist, Auguste Victor Louis Verneuil. He developed an ingenious chalumeau—an upside-down blowpipe—which was immediately successful in making remarkably good ruby. Today, rubies cut from this material are very superior in colour, clarity and size to almost all natural rubies. Chemically and structurally—even under X-ray examination—these gems are almost identical with their natural counterparts.

Other crystal-manufacturing methods use solutions of the substances to be formed. Many minerals do not dissolve well in water. To overcome this

Quartz, one of the commonest minerals, occurs in many varieties. Here are two specimens from Mexico.
FACING: *Sculpture-like agate from Chihuahua.*
ABOVE: *Amethyst from Vera Cruz.*

difficulty, other dissolving agents or fluxes are used at high temperatures, as in the techniques developed by the Bell Telephone Laboratories for growing synthetic ruby. A platinum container about ten inches high and six inches in diameter is filled with a flux of eleven pounds of lead oxide (PbO) and fourteen ounces of boron oxide (B_2O_3). To this is added the material for making the ruby itself: twenty-six ounces of aluminium oxide (Al_2O_3) and just a trace of chromium oxide (Cr_2O_3) to give a red colour. The container and its charge are then heated in an electric furnace to 2400 degrees Fahrenheit, so that the chromium oxide and aluminium oxide dissolve in the now-liquid flux. During six hours of heating at this temperature the contents are stirred continuously by rotating the pot first one way and then the other. After six hours, the tem-

perature is slowly and evenly dropped about 170 degrees Fahrenheit per day for several days. As the temperature drops just below 2260 degrees Fahrenheit, ruby crystals begin to form. When cooling is finished and the whole mass has become solid, with the new ruby crystals mixed in it, the flux is dissolved by soaking in nitric acid, which has no effect on the ruby crystals.

Another useful technique for growing crystals from a high-temperature melt was originated in 1918, but was not developed and used until 1926. Lately it has been used to grow silicon and germanium single crystals for use in transistors. For this purpose, the container is made of temperature-resistant platinum or some other metal, such as iridium or rhodium. It is heated and loaded with the powdered material from which the crystal is made. A small crystal of the metal to be crystallized, perhaps germanium, is suspended over the now melted germanium on the end of a rapidly rotating shaft. It is then lowered carefully and allowed, while still rotating, to touch the surface of the liquid germanium. Still rotating, the original germanium sample is raised very slowly, revealing a new growth upon the seed crystal, a growth that increases in dimension as the shaft rises. It looks almost as if a cylindrical crystal of the metal is being pulled up out of the liquid. Such 'pulled' single crystals can be grown to remarkable lengths. Using this method, Dr. Kurt Nassau of Bell Telephone Laboratories grew a crystal of scheelite ($CaWO_4$) that eventually reached a length of eighteen inches.

Crystals can be grown also from water solution in the laboratory. Usually very high temperatures and pressures are needed to get the various substances into solution. Large quantities of valuable quartz crystals are grown this way for use in the electronics industry. Because water and heat are essential to the process, it is called a hydrothermal method—*hydro* meaning water and *thermal* meaning heat. Equivalent natural quartz of the type that is laboratory-grown is so rare and so expensive that it is economically sounder to manufacture it. Pressures up to fifty thousand pounds per square inch are used at temperatures of as much as 1300 degrees Fahrenheit, and substances are present to help the quartz go into solution.

Under these extreme conditions the autoclaves or 'bombs' containing the solution must be extremely strong. They must have leakproof and blowout-proof closure devices and must be silver- or platinum-lined to prevent contamination of the solution by chemical attack on the lining by the mineralizer. Usually about six inches in diameter and twelve feet long, industrial bombs have four-inch-thick walls that are made of cold-rolled steel.

The bomb is loaded with water, mineralizer and a quantity of small pieces of pure natural quartz in the bottom. At the upper part of the bomb are hung the seed crystals of quartz cut in the form of flat plates. The bomb is sealed and heated electrically to crystallization temperature. The pressure automatically rises tremendously because of the heat-expanded contents. After about twenty days the cycle ends; all the small pieces of natural quartz from the bottom have been dissolved in the water solution and redeposited on the surface of the seed crystals to make fat new crystals of quartz.

If man or nature has provided the proper conditions and sufficient material to make a crystal, a small cluster of atoms will come together and make a starting unit or seed. Sometimes only a few seeds form, and the resulting crystals are large and externally well formed, with distinct crystal faces. If many seeds suddenly form, the result will be an interlocking mass of small crystals

FACING: *Snow Crystals—a reproduction from the rare* Les Pierres Précieuses *by J. P. Rambosson (1870).*
ABOVE: *A high-tridymite structure model representing tridymite that has crystallized at a temperature higher than normal.*

atoms that is repeated many times through space to make up the total lattice. This tiny, three-dimensional unit is called a *unit cell*. It is the shape and symmetry of this unit cell, repeated again and again, that ultimately determines the shape and symmetry of the final crystal. Sometimes the actual shape of the unit cell appears as the outside shape of the crystal. At other times the external faces do not seem related to the shape of the unit cell. Actually, they are always precisely related to it. A crystal face is nothing more than a flat surface made of tremendous numbers of atoms all arranged in the same plane. Since each plane of atoms is a potential crystal face, several different combinations of faces are possible for the crystal. It may not end up, therefore, being exactly the same shape as the unit cell, but may exhibit faces represented by any prominent plane of atoms in its structure.

Today's mineralogist studies space lattices and unit cells by using modern X-ray techniques. X-ray studies tell us a great deal about the possible arrangement of the space lattice in a crystal. However, they do not make absolutely certain the choice among the few possible lattices for a given crystal. Other evidence is needed. The study of crystal morphology is a highly developed science, which explores the external shapes of crystals and the relationships that exist among those shapes.

When a crystal has a *plane of symmetry* it means that an imaginary plane passed through it will divide it into two halves that are exact mirror images of each other. To discover an *axis of symmetry*, a crystal is rotated or spun on an imaginary axis passed through it. If, during a complete turn on this axis, the crystal assumes at least one position that makes it look identical with the way it started out, then an axis of symmetry is present. This kind of crystal symmetry can occur two, three, four, or six

showing no particular external characteristics. Such a mass is considered crystalline, but not a crystal.

As the atoms of a crystal begin to cluster, they are forced into positions related to each other in a highly organized manner. The positions taken depend mostly upon two things—the size of the atoms and the amount of attraction they have for each other. The atoms build themselves up into three-dimensional networks of repeating patterns, some of which are quite simple and others very complex. The ways in which the patterns repeat are called the space lattices and, as Bravais had predicted, there are only fourteen different lattices possible. All others are merely variations of the fourteen arrangements.

A close look at the space lattice of any crystal shows that it consists of a unit arrangement of

Examples of well-crystallized minerals. ABOVE: *Exceptional orthorhombic crystal of sulphur from Agrigento, Sicily.* FACING, TOP: *Iron-stained hexagonal calcite, Cumberland, England.* RIGHT: *Cockscomb colony of small orthorhombic marcasite crystals, Picher, Oklahoma.*

times for one turn on the axis, depending on the species of the crystal. A crystal is symmetrical about its theoretical centre—or has a *centre of symmetry*—if any line drawn through this centre point always intersects the opposite faces of the crystal at identical distances from the centre point.

The detailed measurement of thousands of crystals with improved instruments has resulted in the determination of just which planes, axes and centres exist together, in combination, in natural and synthetic crystals. All crystals can be conveniently and logically grouped or classified into only thirty-two different kinds of symmetry, or crystal classes. For the sake of further convenience and logic, these thirty-two classes are grouped into six crystal systems, all classes in a given system having some important symmetry in common.

The systems are: isometric, tetragonal, ortho-rhombic, monoclinic, triclinic and hexagonal. Each system is recognizable by a set of unique symmetry axes assigned to it and by means of which its symmetry family can be easily described. On page 47, block models of typical crystals in each of these six systems are pictured, with diagrams describing their mathematical reference axes.

A set of identical faces on a crystal is called a form. Everyone recognizes the cube, a form belonging to the isometric system. It consists of a set of six faces, each one pierced at right angles by a crystal axis and each one exactly like the others. Another form in the isometric system is the octahedron. This form consists of eight faces, each an identical equilateral triangle. The faces are positioned so that the octahedron looks like a double-ended square pyramid.

Fighting the multitude during the rush hour on the London underground or boarding a Japanese train with the help of an official 'pusher' can give some insight into the problems an atom has in occupying its place within a crystal lattice. When the rush starts and the seats begin to fill, some of the seats and some of the standing spaces are improperly used: two people are crowded in where only one fits, or perhaps someone is occupying a space and a half. The result is a dislocation of proper packing.

Since the growing of a typical crystal requires the proper placement of about sixteen trillion atoms an hour, the number of defects in the packing tends to be phenomenal. But even more phenomenal is the number of atoms that do manage to get into the right places. A normal accumulation of defects resulting from the organization of a crystal is only about one-hundredth of one per cent—far better than the underground manages.

One way to reduce defects in man-made crystals is to slow down drastically the rate of packing. But even if the rate is slowed to one-tenth of normal, more than one and a half trillion atoms must still be packed in the crystal in an hour. A better method is to slow down the packing rate, or crystal growth, just enough to see what causes the defects in placement, thereby learning to change the conditions so as to avoid creating these defects. This is one method currently in use for studies of crystal growth. This slowdown can be accomplished by using solutions with less material in them or by controlling the rate of temperature reduction. But whatever the rate of growth, crystals are always subject to problems that cause structural disorder.

One convenient way to look at the various disorders in crystals is to arrange them by point disorders, line disorders, plane disorders, and solid or three-dimensional disorders.

By *point disorders* we mean that certain points in the space lattice are not properly occupied. Quite often, the liquids or gases from which crystals

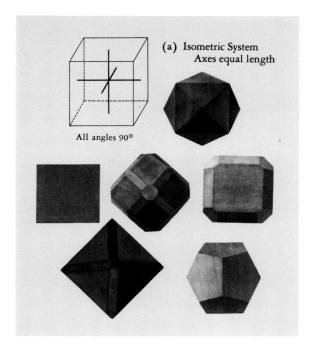

(a) Isometric System
Axes equal length

All angles 90°

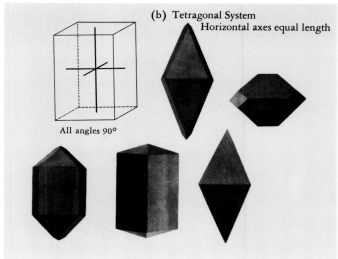

(b) Tetragonal System
Horizontal axes equal length

All angles 90°

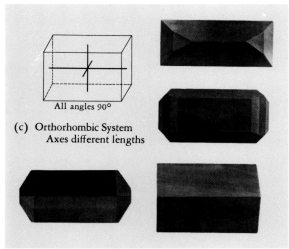

All angles 90°

(c) Orthorhombic System
Axes different lengths

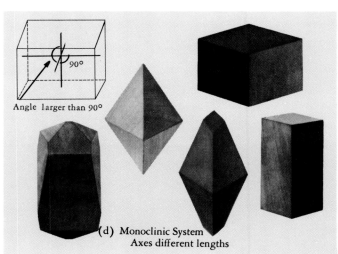

Angle larger than 90°

90°

(d) Monoclinic System
Axes different lengths

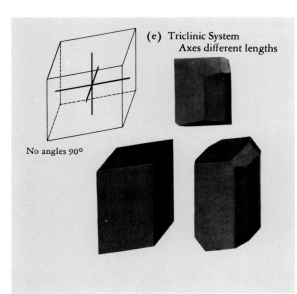

(e) Triclinic System
Axes different lengths

No angles 90°

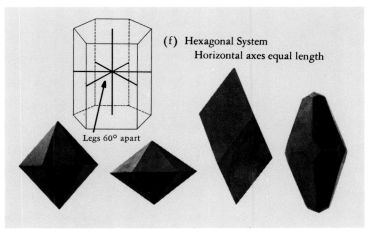

(f) Hexagonal System
Horizontal axes equal length

Legs 60° apart

*Crystal-forming atoms organize in one
of the six crystal systems shown,
with characteristic reference axes. Typical
minerals in each crystal system are:
(a) pyrite, fluorite; (b) zircon, rutile; (c) baryte,
sulphur; (d) sphene, gypsum; (e) axinite,
rhodonite; (f) quartz, calcite.*

come are impure, and the atoms of the impurity happen to be about the same size as the atoms that will form the crystal, thereby being able to occupy positions in the lattice normally used by the proper atoms. The structure thus develops normally, but the final crystal is of mixed composition. The mineralogist refers to this as a mixed crystal, or a solid solution. Mixing can proceed to a remarkable extent if the sizes of the atoms do not differ more than 15 per cent.

One of the best examples of mixing occurs in the mineral species fayalite and forsterite. Fayalite is an iron silicate and forsterite a magnesium silicate. They have closely related structures. In nature, a complete series of their combined crystals can be found with almost any proportions of iron and magnesium. Of course, this mixing causes changes in characteristics, and in the case of the fayalite-forsterite series a crude separation can be made just on the basis of colour. Crystals of compositions toward the forsterite or magnesium end of the series are light-coloured greenish-brown, green, or even white. Toward the fayalite or iron end of the series the crystals are dark greenish-brown, brown or black.

Point disorder can be of another type. Crystals, unlike living objects, do all their growing by adding new material in layers of atoms on the outside of the crystal. This is called growth by accretion. In the hustle and bustle of trying to get hundreds of layers of trillions of atoms quickly and accurately set in place, it is very easy for some new layers to be deposited before all the lattice positions underneath are fully occupied. This means that many atom positions in a crystal are really occupied by a ball of nothing. These empty spots are called voids. An excellent example of this is the mineral pyrrhotite, a combination of iron and sulphur. When analyzed chemically, this mineral was described as solid solution or mixed crystal, with an excess of sulphur in the normal iron sulphide. It turned out, however, that there was not too much sulphur but too little iron. Many of the positions in the lattice that should be occupied by iron atoms were vacant. The realization that many such minerals have voids in their structures has helped to explain a number of puzzling chemical compositions.

There is still another kind of point disorder which is related to that found in mixed crystals. Sometimes the impurity atoms may be very much smaller than those in the lattice. If so, they can easily slip into spaces among the atoms of the lattice, which is such an open network that there is plenty of room between occupied spaces. Metal crystals very often take up atoms of hydrogen, carbon, boron or nitrogen—all small atoms—in these spaces. In structures of the amphibole family of minerals there are holes in the network that are large enough to take in sodium atoms.

Plane disorders are another common occurrence. Mention has already been made of crystalline masses, made up of groups of crystals that form together as touching grains rather than as free-standing individuals. There is often disorder in the planes of atoms that are located at the boundaries where the grains touch.

A grain boundary appears to be a no-man's-land between two crystal grains. The problem is that atomic planes do exist at the boundary, but the atoms in these planes have a difficult time deciding which crystal's structure they should be lined up with. Some of the atoms align themselves with one crystal and some with the other, so that for several layers there is disorder.

Even with a freestanding, well-developed single

FACING: *Pyrolusite, a black manganese oxide, sometimes grows as branching crystal dendrites embedded in sedimentary rocks. This remarkably fern-like sample is in shale from Solenhofen, Germany.*

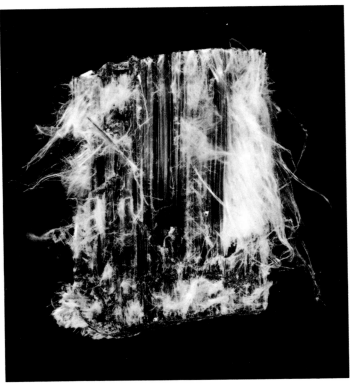

crystal, there is often the same sort of disorder on the outside layers. Atoms in these layers are in a different situation from that of their colleagues at the interior of the crystal. In the interior all the atoms are totally surrounded by others. On the surface the atoms have no layers over them and are, therefore, relatively more free. Since much of the influence of nearby atoms is missing, these surface atoms are slightly more active and will be forced into somewhat different positions. They are also more susceptible to chemical activity and to removal from the surface of the crystal. Every time a crystal breaks, in fact, some of its interior layers are exposed. The atoms in these newly exposed layers must now shift positions and make other adjustments to fill a role somewhat disordered from the pattern of the rest of the structure.

Line disorders caused by improper placement of lines of atoms, rather than points or planes, have great effect on the growth of crystals. When extra lines of atoms are trapped in a structure in its pell-mell building rush, they obviously must force it out of alignment. It may be many atom layers later before the structure is able to compensate to straighten out the ripples. Since there are so many of these dislocations, some manage to cancel each other out, while others tend to add their effects and make the situation worse. An accumulation of dislocations can stress and weaken severely the structure of a crystal.

Strangely enough, surface dislocations on a crystal can encourage it to grow more easily and more rapidly. This is because an atom, when it prepares to become part of a crystal structure, may attempt to attach itself to an already formed layer of atoms. Of all the possible attachment sites this is the least desirable, because the only attractions pulling it in come from atoms in this layer. A

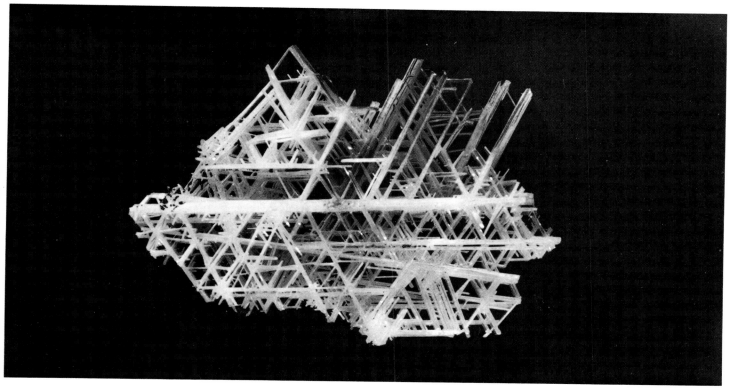

much better site would be on the surface of one layer of atoms and up against the edge of another partially formed layer sitting on top. In this position the new atom is attracted not only by the atoms in the layer under it, but also by the atoms at the edge of the layer alongside. Such a place is known as a preferred deposition site.

The presence of a dislocation, or extra row of atoms in the crystal, can force part of the surface up and out of line with the rest, producing what is technically called a screw dislocation. Such a screw dislocation supplies a ready-made preferred deposition site, encouraging the attachment of atoms and accelerating growth. The earmark of such screw dislocation growth is the spiral trace that it leaves on the crystal face in the form of a set of steps.

Line disorders, or dislocations, can exert their influence in a few or many directions in the disordered structure. In some cases such dislocations are uniformly distributed in all directions and growth proceeds with ease in all three dimensions. Some crystals with dislocations encouraging growth in only two dimensions tend to grow as flat plates. Some crystals develop dislocations encouraging growth in only one dimension and thus have little encouragement to grow in width and thickness. They are called whisker crystals and grow rapidly in length—often hundreds of times longer than they are thick.

Disorder in three dimensions would mean that atoms were out of position in all directions. *Three-dimensional disorder* is another way of saying that the material may appear to be solid but does not meet the requirement of an orderly internal structure. The description of glass, given earlier, offers a perfect example of such disorder.

Because of all the many internal stresses and strains, it is a wonder that crystals form at all.

Growth habits and impurities, among other factors, influence crystal form. FACING, FAR LEFT: *Asbestos—fibrous habit.* LEFT: *Sand calcite with impurities.* ABOVE: *Net-like growth habit results in reticulated cerussite.*

51

*Accidents of crystal growth:
(a) growth hillocks on diamond
crystal; (b) oriented hematite
on quartz; (c) striations on
anatase; (d) silver dendrite
of oriented octahedra;*

*(e) 'Japanese' twinning in
quartz; (f) hopper cube
faces on diamond crystal.*
FACING: *Radiating aggregates
of cyanotrichite crystals.*

When they do form, it is understandable that their appearance will reflect the interplay of conflicting or cooperating internal events. The irregularities in crystal growth are important scientifically because of the clues they give to the secrets of crystal growth. And for the hobbyist and collector they add considerably to the beauty or oddity of crystal specimens they have acquired.

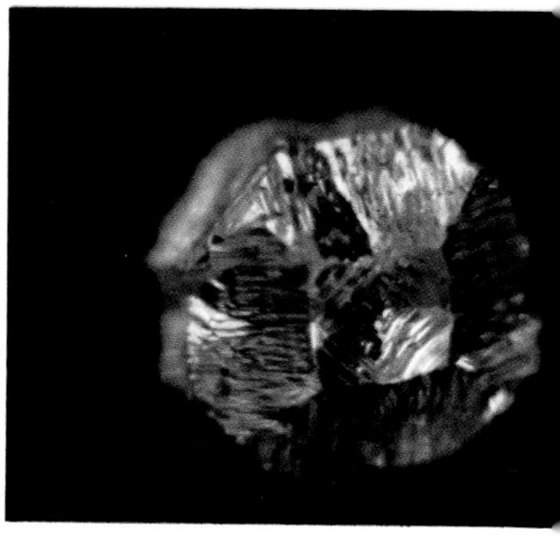

Along with all their other difficulties in growing, crystals have a tendency to become malformed and to deviate in shape from a textbook ideal. If the malformation expresses itself in a failure to form recognizable, flat crystal faces—even when there

is good internal structure—the crystal is said to be anhedral, *an* meaning not and *hedral* meaning formed. Should the crystal have faces that are poorly formed, it is subhedral—less than formed. If the faces are excellent the crystal is euhedral, *eu* meaning well.

The whole problem of malformation arises because something has favoured the development of one group of faces and not another. One of the known causes is a steady flowing movement of the solution while the crystals are forming. As the solution flows through the crystal-lined vein or cavity, all the crystal parts facing into the flow

receive a fresh supply of building materials and grow rapidly. Those parts in the lee receive only the material that eddies around to them behind the crystal. Although they also grow, it is a slower process.

We must remember that even when the crystals are badly formed, there is an underlying organization. The crystal structure stays the same. All the crystal faces, be they too large or too small, are still at the same angles to each other, and Steno's old Law of Interfacial Angles remains valid.

Miscellaneous designs and figures, standing out in very low relief on an otherwise perfectly smooth and glassy crystal face, are called growth hillocks or growth figures. Sometimes these appear as flat planes or shallow curved surfaces jutting out at very slight angles from the face and covering a considerable portion of it. In such cases they are called vicinal faces.

Many of the better-defined hillocks seem to come from the spiral growth resulting from screw dislocations. When two such dislocations of opposing direction are near neighbours, their respective growth plateaus sweep outward and eventually meet, creating a single plateau which continues to grow toward the edges of the crystal. Since the

opposing dislocations are still there, their spiral growth continues to sweep on around, creating a new cycle on a new plateau on top of the original. This continues, plateau upon plateau, until a small hillock is built. If crystal growth is arrested before this pile of tiny plateaus can grow sideways toward the edges of the crystal, it remains as a small, isolated projection—a little hillock—on the main crystal face.

Sometimes the forms of the hillocks show no particularly definite outline. At other times the geometrical outlines they do assume are good evidence of the internal structure of the crystal. Because of this, a rough, broken crystal fragment with at least one good face showing growth figures can often be identified, and the direction of its structural planes determined, without recourse to X-ray techniques. The growth hillocks on crystal faces sometimes continue their growth and become larger single crystals in their own right. Since these hillocks have begun as structural continuations of

the parent crystal, this development remains perfectly lined up with the structure of the original crystal. Thus when the parent crystal is rotated under a light, all the related faces of all the individual crystals grown from the various hillocks will flash their reflections in unison. Because all the parts are parallel to each other, this is called a parallel overgrowth.

Parallelism of this kind can also be caused by certain kinds of coatings on a crystal during its growth. A parent crystal may stop growing and become coated with clay, or hematite, or some other mineral. When this coating is thin at certain spots, the original crystal may begin to grow again. Growth is arrested elsewhere by the thicker coating, but each thinly covered spot grows a new crystal portion parallel to each of the others, because all are actually continuations of the crystal's original structure.

There is another striking kind of parallel overgrowth that consists of a crystal knob growing on the end of another crystal, so that it resembles a king's sceptre. Quartz supplies the best examples of this sceptre growth which, for some unknown reason, limits itself to one large individual growth rather than producing a number of small ones—and that one only at one end of the parent crystal.

The concept of an un-growing crystal has an Alice-in-Wonderland sound. Nevertheless, the process that added atoms to the structure as they lost some of their energy can be put in reverse when energy is added to the atoms that are already part of the structure. This is called dissolution, or the return of atoms to solution, and it actually is a process of un-growing that occurs in nature. It starts at random spots on a crystal face and spreads from each spot, gradually leaving a series of small, shallow areas called etch pits. If this etching process is carried on long enough, the entire crystal will go. If arrested, however, the pits can be studied as evidence of crystal structure. Just as crystal growth is precisely controlled by its structure, so is etching.

In nature the etching process starts when the environment that gave birth to the crystal changes to conditions under which the crystal cannot exist. Sometimes, as the crystal starts to go, there are so many tiny etch pits that to the naked eye the crystal face may look frosted, or as if grinding has removed its shiny surface. Magnification, however, will reveal the myriad tiny pits, and show that each of them has the normal triangular, rectangular or other geometric outline typical of the crystal structure.

There are a number of minerals that occur in curved, twisted or bent crystals. The causes for this kind of growth have not been completely determined. Whatever the reasons, the crystals themselves are unusual because they introduce curved surfaces into a world of flat planes and

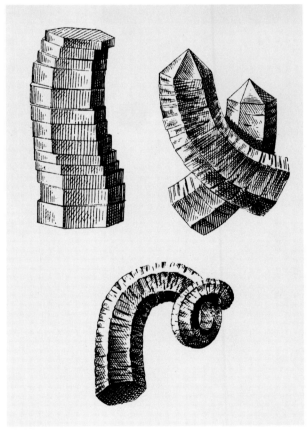

More crystal growth accidents. FACING: *Hoppers of halite —salt—from California.* TOP: *Parallel crystals of galena from Joplin, Missouri.* ABOVE: *Curved stacks of chlorite crystals, from* Corso di Mineralogia *(1862).*

geometric figures with measurable angles. The Swiss *strahlers*, who are professional crystal collectors, are always searching for *gwindeln*—twisted crystals of a smoky-coloured quartz, found in Alpine veins among more normal quartz and other fine mineral specimens. These twisted quartz crystals are actually a series of flattened quartz crystals stacked up on their sides, each one slightly turned from its neighbour. Sometimes the individuals in the stack are so small that when seen together with the naked eye the whole looks like a single curved crystal.

Dolomite crystals are often stacked together in little groups—again with each crystal slightly turned —that look like small, cream-coloured saddles. Stranger still are some crystals of mimetite, especially a yellow to orange-brown variety called campylite, which occurs in England. Many of these individual hexagonal crystals are shaped so that they resemble

tiny barrels with both ends flat and with curved edges swelling out at the crystals' middle.

Strangest of all are tiny crystals of a mineral called heteromorphite, from Rumania. Normally this mineral, a compound of lead, antimony and sulphur ($Pb_7Sb_8S_{19}$), occurs in long, shiny, metallic, hair-like crystals. Now and then, however, a few appear that resemble platinum wedding rings. They are completely circular and almost the colour of platinum.

Crystals of a single mineral species are the product of only part of the activity that has occurred within their parent solutions or melts. These solutions and melts also contain atoms of other elements waiting for the proper change in conditions so that they too can form crystals. When they do finally crystallize, there is a good possibility they will settle themselves on the crystals that have already formed. Thus they become coatings of one kind or another.

Some coatings are very selective about the sites upon which they will deposit themselves. In the quarries around Paterson and Prospect Park, New Jersey, quartz crystals are found with normal six-sided pyramid terminations. Some of these are coated with a bright red hematite film. Oddly, all six faces are not coated—only alternate ones— creating a gay pinwheel effect. The hematite has recognized that the structure underlying three of the faces is compatible with its own structure, whereas that under the other three is not, and has selected accordingly. The six-sided pyramid is not as simple a structure as it seems to be on first reflection.

Sometimes when one mineral is deposited as a coating on another, its crystals are big enough to be seen as separate entities. Surprisingly, they may be perfectly lined up in one definite structural direction of the crystal host. Although the structures

Three excellent crystal samples of their species.
FACING: *Twinned tetrahedrite with chalcopyrite, from*
Harz Mountains, Germany. TOP: *Wulfenite from*
Chihuahua, Mexico. ABOVE: *Celestine from Chittenango*
Falls, New York.

of the two minerals are not alike and cannot form solid solutions, they must have something in common. The outer layer of the host crystal is enough like some atom layer in the overgrowth species to confuse the overgrowth into accepting the host layer as one of its own. The overgrowth crystal then proceeds to build its own structure. Perfect examples of this are red rutile crystals, which grow in a flat, net-like pattern on black hematite plates; gold-coloured chalcopyrite crystals, which look like thousands of tiny pyramids lined up on the surface of red-brown sphalerite crystals; and tiny bright blue pseudoboleite crystals, which stack like small boxes, one each on the six faces of a cube of blue boleite. Coatings like these are known as oriented crystal overgrowths.

Nature often produces what is called phantom crystals. These are ghostly, crystal-outline images appearing inside a transparent, glassy crystal. The phantoms are only the remnants of crystal coatings that occurred during earlier growth stages. Often, as a crystal forms, its growth stops long enough for the surface to become coated with a variety of mineral substances. If this coating is too heavy, growth stops completely. If, however, the coating is only a light dusting, the crystal can begin to grow again when material is available and conditions are right. This traps the coating in the new growth, a process that can repeat itself again and again so that multiple phantoms, one inside the other, may be formed.

Conditions under which a crystal is growing may become so unfavourable that there is often some question about whether the process will continue. In such a case, growth will most likely proceed along all the edges of the crystal and not on its faces. This produces a kind of skeletal, open-network crystal with all edges and no faces.

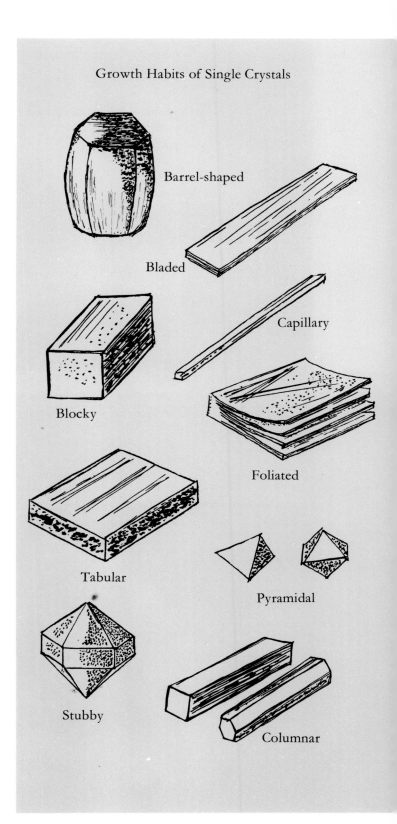

Growth Habits of Single Crystals

Barrel-shaped

Bladed

Capillary

Blocky

Foliated

Tabular

Pyramidal

Stubby

Columnar

In different colonies, aragonite appears to be three different minerals. FACING, TOP: *Acicular aragonite from Utah.* FAR LEFT: *Coralloidal aragonite, Styria, Austria.* LEFT: *Pisolitic aragonite from Bohemia, Czechoslovakia.* ABOVE: *A sampling of shapes—growth habits—of single crystals.*

Such crystals are called hoppers.

There are various reasons for this kind of growth behaviour. One has to do with the fact that when an atom joins a crystal structure it gives off its excess energy. This energy must be removed from the crystal before other atoms can be deposited. Physicists have demonstrated that energy leaks off best from objects at sharp points or edges. That is why lightning rods are usually pointed. Since energy leaks off best from crystal edges, rather than faces, edges become the best areas for continued crystal growth. Whatever the explanation, the fact is that whenever growth on the faces of a crystal is suppressed it often continues along the edges, thereby producing these skeletal hoppers.

When rotated on a symmetry axis, a cube of pyrite at first seems to show a fourfold repetition of identical faces as a true cube should. This is very high symmetry and is obvious for true cubes, but a pyrite cube is deceptive. A close look at this pyrite cube will reveal that each face is covered with hundreds of parallel ridges or channels, which run in different directions on different faces. These are called striations, or striae.

These striae are the result of a battle between two different crystal faces, the cube and the pyritohedron, trying to form in the same place at the same time—a process called oscillatory growth. As each battling face momentarily dominates the growth cycle, it leaves a little flat strip of face running in its own direction, only to be cut off by a flat strip running at a different slope. In the example discussed here, the cube has ultimately dominated, so that the crystal looks pretty much like a true cube, except that the grooves on its faces run in different directions on neighbouring faces. Rotation of this crystal will not give four identical repetitions because the striations do not match directions. It has no fourfold symmetry.

Pyrite has a different symmetry, then, from a mineral like fluorite, whose cubes always show four exact repetitions when rotated. The presence of striations on crystals, therefore, helps to determine the correct symmetry for pyrite as well as many other minerals.

While it is not a constant characteristic, each mineral tends to form in crystals of a particular shape. For instance, fluorite crystals can occur with different sets of faces—the cube, octahedron, dodecahedron, and several others. Almost always, however, they occur as cubes. The habit, or usual shape, for fluorite, then, is the cube. This means that the fluorite structure favours a process of growth producing cube faces.

Strangely enough, crystals end up favouring the faces that grow most slowly. The cube faces, for example, are the kind that grow slowest in fluorite. This situation can be disrupted and crystal habit can be changed by the introduction of impurities. The impurities tend to congregate on the faces where most of the growth activity is found. They interfere with the growth of these faces, slow them down and give them a chance to compete with the slowest-growing crystal faces, which usually dominate the final shape of the crystal.

Some easily recognized overall shapes, or growth habits, of crystals are shown on page 59.

Acicular: From a Latin word meaning needle, this refers to crystals that are long and thin, like those found in the species natrolite and mesolite.

Barrel-Shaped: Self-descriptive term applied to such crystals as vanadinite and mimetite.

Bladed: Crystals that are flat, broad and long, like a knife blade, are so described. Kyanite and hemimorphite are good examples.

Blocky: This describes the appearance of crystals

FACING, TOP: *In sedimentary agate from Arizona, crystals are fine-grained and undetectable.*
LEFT: *Igneous dark red almandine garnet from Spruce Pine, North Carolina; crystals are walnut-size.*

that are almost of equal dimension in all directions and look like children's toy blocks. Some of the feldspars, galena, and fluorite can be so described.

Columnar: These crystals are thick and fairly elongated, shaped in miniature like the columns of a building. Examples include beryl, quartz and tourmaline. Sometimes the word *prismatic* is used to describe the same crystals, because the dominant faces on columnar crystals are usually called prisms.

Capillary: From a Latin word meaning hair, capillary is often used interchangeably with *filiform*, also from the Latin, and means thread-shaped. The terms describe minerals such as silver, which is sometimes found as thin wires, or millerite, which is found in long, very thin, stiff wires.

Pyramidal: This expression is used when the ends, or even the entire crystal, look vaguely like three-, four-, or six-sided pyramids. Examples include wulfenite and anatase.

Stubby: Such crystals are also sometimes described as *stout* or *equant*. All three terms try to draw an image of crystals that are neither flattened nor elongated, but are nearly the same dimensions in all directions. Tourmaline, apatite and beryl can be columnar or prismatic, but they may also be of shorter lengths and stubby.

Tabular: This means that the crystal's growth has been flat, and baryte is a common example. When very thin in relation to their area, crystals—such as torbernite and wulfenite often form—are called *platy*.

Crystals seem to prefer to collect in groups. Sometimes they have little choice, but the result is that an isolated, single crystal is a relatively rare thing in nature. Most rocks and minerals of the earth's crust are in aggregates of individuals that are anhedral—with no distinguishable crystal faces. These aggregates vary from those so tiny that they are just visible under a good optical microscope

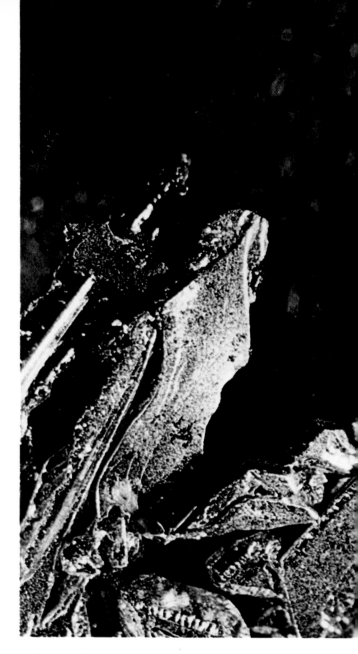

to those so large that individual crystals in the group are several feet long. Almost always the arrangement of crystals in their aggregates is haphazard. Only rarely are they orientated toward each other in some particular way.

In the early days of mineralogy, differences in the kinds of crystal groupings were given undue importance. Some minerals were even divided into several species because their crystals looked different in different groupings. Following are some of the most easily recognized crystal groupings:

Arborescent: These are tree-like groups, with the

Native gold occurs fairly often in branch-like dendritic groups, like this handsome gold dendrite from Transylvania, Rumania. The colour can vary considerably, generally depending on the amount of silver present.

crystals in branching, perhaps fern-like, arrangements. These are sometimes called *dendritic* groups from the Greek word for tree. More species tend to crystallize as dendrites than in any other way. Natural gold, silver and copper crystals generally form this way. Pyrolusite and psilomelane, two oxides of manganese in dendrites, account for the beautiful black, fern-like staining that appears in many sedimentary rocks.

Amygdaloidal: From a Greek word meaning almond, this refers to almond-shaped cavities that are commonly found in volcanic rocks. These cavities were originally the same spherical shape as the gas bubbles that formed them, but the flow of the rock flattened and stretched them. Often they are filled with collections of minerals, phillipsite, analcime, heulandite or others, which then become known as amygdaloidal minerals.

Bladed: Individual crystal blades are sometimes massed together in aggregates; this term can describe either the separate crystals or the group.

Colloform: There is a tendency to use this term to replace several others that mean roughly the same thing. All colloform aggregates are partially

spherical or rounded on the outside. *Botryoidal*, from the Greek, means that the aggregate looks like a cluster of grapes. *Mammillary*, from the Latin, means that it looks rounded, like a breast. *Reniform*, from the Latin for kidney, again refers to a roughly rounded form. *Globular* and *spherulitic* are self-explanatory, describing a rounded shape.

None of these expressions refers to the shape of individual crystals in the group, but only to its total rounded form. Malachite, hematite and wavellite all seem to grow from a number of acicular or needle-shaped crystals deposited on a surface. As they continue to grow, those that push outward from the surface have access to fresh material and continue their growth, while those lying in other directions remain stunted and are eventually covered up by the rest. Thus the aggregate becomes a group of thickening needles pushing out from the surface, and eventually growing together to form a mass with a rounded surface composed of the ends of the crystals. The term colloform is used because it is believed that most crystal aggregates of this kind grow from gelatin-like accumulations of the mineral material. These are called colloids, from a Greek word meaning glue.

Compact or massive: These terms refer to small,

interlocked crystals closely packed, so that they do not appear to be crystallized at all. Some of the clay minerals are good examples of this kind of grouping.

Cockscomb: Flat fan-like arrangements of crystals, with their pointed ends giving a sawtooth effect, are commonly found in the mineral species marcasite.

Coralloidal: The twisted, tangled, cylindrical growth of aragonite, which makes it look like a mass of white coral, gave rise to this term.

Drusy: A mineral or rock surface coated with a layer of fairly uniform crystals of small size is called a druse.

Earthy: A mass that crumbles like dried mud and seems to have no crystalline character is earthy.

Fibrous: Asbestos is a typical fibrous mineral. It consists of very compact bundles of long, fibrous, almost silky crystals. It is usually easy to strip off small groups of silky hairs from a fibrous mass.

Granular: This refers to a mass made of small individual crystal grains.

Micaceous: This word describes the characteristic of mica minerals that permits them to be split off easily in very thin sheets. Sometimes *lamellar*, from the Latin for plate or layer, and *foliated*, from Latin for leaf, are used to describe this characteristic.

Nodular: The term is used to describe lumps or kernels or nodules of the mineral, but does not indicate whether the object in question is a single crystal fragment or a multiple crystalline mass.

Oolitic, Pisolitic: Both of these terms are used for small concentric, globular masses of mineral. Oolitic is used for globules about the size of fish eggs, and pisolitic for objects as big as peas. Anything larger is usually just called a concretion.

Plumose: Sometimes the mica minerals and their near relatives form in scaly, spreading accumulations that resemble feathery plumes.

Radiated, divergent: The crystals in this group

64

Crystal Colonies
Growth Habits of Groups of Crystals

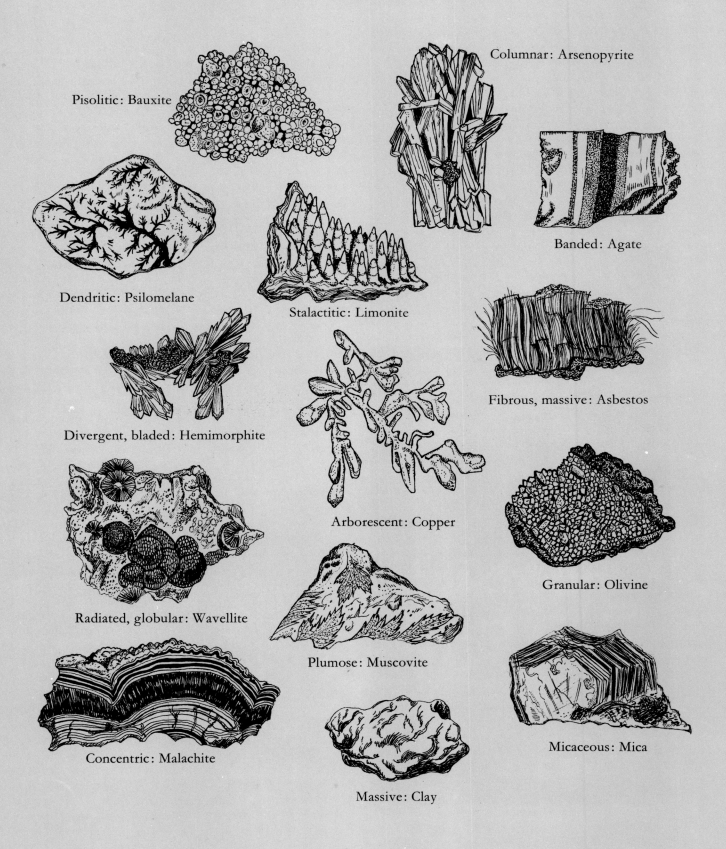

Pisolitic: Bauxite

Columnar: Arsenopyrite

Banded: Agate

Dendritic: Psilomelane

Stalactitic: Limonite

Divergent, bladed: Hemimorphite

Fibrous, massive: Asbestos

Arborescent: Copper

Granular: Olivine

Radiated, globular: Wavellite

Plumose: Muscovite

Concentric: Malachite

Massive: Clay

Micaceous: Mica

start growing at some central spot and spread outward. If they separate from each other enough to be distinguished individually, they are divergent.

Reticulated: This word comes from the Latin and means net. It refers to the open lattice made by numerous crossed bars of some crystallized minerals. Reticulated cerussite, reticulated rutile and reticulated cuprite (illustrated on the left) are good examples.

Rosette: Any large mineral collection contains at least one example of Swiss *eisenrose*—literally, iron rose. This is a crystal group of flat, shiny hematite crystals arranged in an overlapping pattern around a centre—just as are the petals of a rose.

Stalactitic: The Greek origin of this word refers to the oozing out of drops necessary to produce cave formations known as stalactites. Any deposit of mineral material formed by the same process—such as malachite, aragonite, goethite or calcite—is said to be stalactitic.

Ordinarily crystal aggregates are accidentally arranged, or at best organized in a crude way. There are many groups, however, of two or more crystals that are joined to each other as are Siamese twins, sharing the same structures. These twins start to form early in crystal growth when some of the atoms, instead of occupying proper lattice positions, find themselves in other positions which the energy of the lattice permits but does not prefer. These atoms get there because of the mixups that occur during rapid crystallization. Once there, they form a kind of nucleus from which crystal growth continues. The two segments then continue to grow side by side, quite compatible because they are structurally comfortable together. The plane at which the two parts of the twin join is called the composition plane. Usually it is not difficult to see where twins join, because at the intersection the

ABOVE: *Cuprite, a red copper oxide, sometimes occurs in whisker crystals, which are really grossly elongated cubes.* BELOW: *Limestone stalactites, Luray Caverns, Virginia.* FACING: *Native copper from Lake Michigan's great Keeweenaw Peninsula deposits.*

crystal faces meet in V-shaped junctions. These are technically known as re-entrant angles because the V's point into the body of the crystal.

Since there are few structural positions that are permitted to the misplaced atom, there are very few angles at which the twins can form. Each known angle is called a twin law. As an example, the mineral staurolite is often found in a twin formation in which the two related parts penetrate each other at right angles to make a cross. This right-angle twin is one of two possibilities for staurolite. Its other twin law permits the crystals to penetrate at about 60 degrees, rather than 90 degrees. For other minerals there are Brazil twins, butterfly twins, Dauphine twins, spinel twins, fishtail twins, bishop's mitre twins.

There are several fundamental kinds of twinning. Twinning with two individuals in contact with each other as if they had been glued together is appropriately called contact twinning. The staurolite twins mentioned above are called penetration twins, be-

cause they look as if one of the crystals had been forced through the other and out the opposite side. Sometimes the contact or penetration twins continue in a circle until a wheel-shaped group is produced. These are called cyclic twins. At other times the twinning is repeated back and forth a great number of times in a row, so that individual twin segments cannot be easily distinguished. Under magnification, however, it can be seen that the structure is zig-zagging back and forth. This is called polysynthetic twinning; it is very common with certain feldspar minerals, such as albite.

For the crystallographer, twinning confuses and complicates the study of crystal symmetry, since it often makes crystals appear to be more symmetrical than they really are. For the mineral collector, on the other hand, the phenomenon produces a great number of geometrically interesting and sometimes beautiful objects.

All statements, definitions and discussions of crystals are rather firm in their insistence on an organized structure. Strangely enough, there are some true crystals that have no organized structure. Perhaps it would be better to say that they had structure at one time, but that it has become thoroughly disordered. Minerals of this kind have *metamict* crystals. Interestingly, they are always radioactive, containing less than 1 per cent of uranium or thorium atoms. This is the direct cause of their strange condition. The original structures, through long periods of radioactive bombardment from the uranium and thorium, literally have been battered into disorder.

It is quite possible to reproduce this metamict condition in the laboratory. The crystals retain the general outward appearance and the same chemical analysis. However, they lose their characteristics of cleavage, behaviour with light, reaction to X-rays, and all others that depend on structure. When some metamict crystals are heated, giving the atoms a new freedom of movement, they regain their crystalline characteristics and give off so much heat that they become incandescent and glow brightly. When we remember that organizing (or reorganizing) a crystal structure means a release of energy for all the atoms, we can understand where the heat comes from.

Crystals develop in so many different ways, and are subject to so many vagaries of growth, that John Ruskin was moved to write of them, 'They are wonderfully like human creatures . . . never think what is to happen tomorrow. . . . You will see crowds of them forced to constitute themselves in a hurry. . . . Then you will find indulged crystals who have changed their minds and ways continually . . . been tired, and taken heart . . . been sick, and got well . . . thought they would try a different diet, and then thought better of it; and made but a poor use of their advantages, after all.'

FACING, FAR LEFT: *Quartz crystals from Cornwall.*
LEFT: *Twinned cassiterite from Bohemia and model.*
ABOVE: *Tetragonal rutile crystal from Brazil is so malformed by twinning that it resembles the orthorhombic model beside it.*

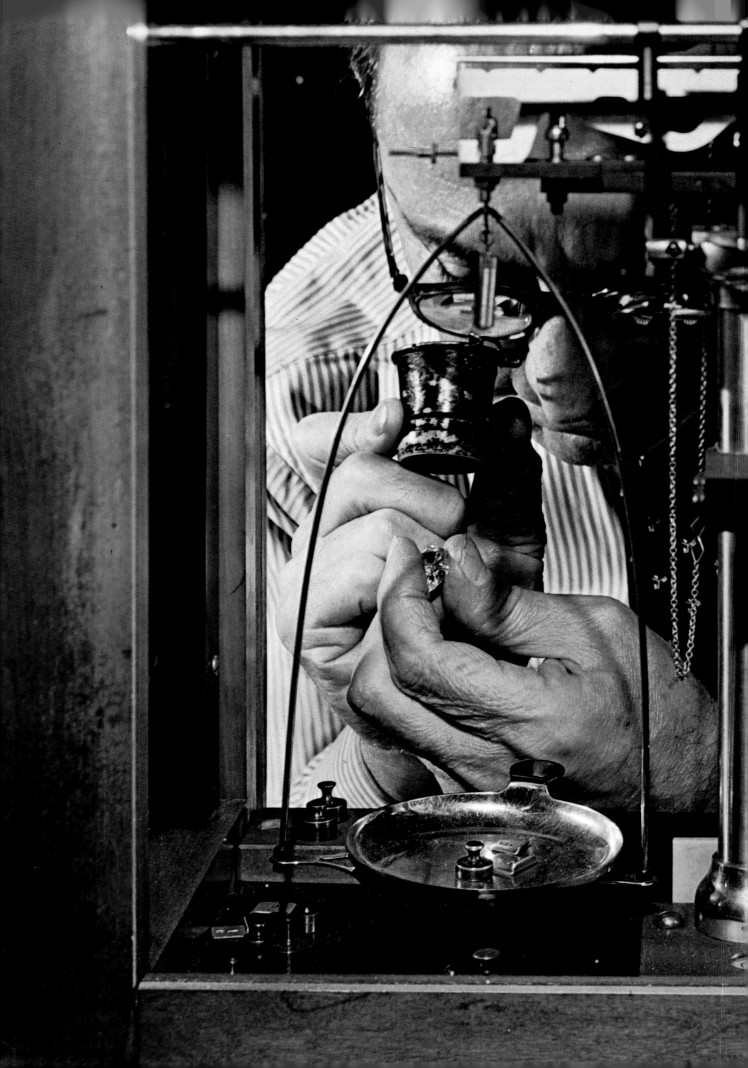

Gemstones: the Royal Line

3

Before the dawn of history men were finding and assigning special values to certain kinds of rock and mineral pebbles that were rare or particularly beautiful. Thus was born the gemstone: a mineral set apart from the rest by beauty, rarity and durability. This combination of attributes is relatively uncommon. Perhaps only one hundred mineral species are considered gems. For the sake of these baubles, wars have been fought, kingdoms have changed hands, and assassinations have been committed.

Gemstone lore, legend and superstition have been part of every human society. In almost every culture gem materials have been given religious or spiritual significance. The signs of the zodiac, for example, have been associated in several religions with gemstones that traditionally helped them to exercise their influence over mortals.

At the beginning of the fifth century, St. Jerome spoke of a relationship between the twelve gemstones in the breastplate of the Jewish High Priest, the twelve months of the year, and the twelve signs of the zodiac. However, the custom of wearing a birth-month stone is not ancient; it does not seem to have developed until the eighteenth century in Poland, and is currently perpetuated in the official birthstone list of the jewellery industry: for January, garnet; February, amethyst; March, aquamarine or bloodstone; April, diamond; May, emerald; June, moonstone or pearl; July, ruby; August, peridot or sardonyx; September, sapphire; October, opal or tourmaline; November, topaz or citrine; December, turquoise or lapis lazuli.

The occult powers of gems may be very much open to question, but they have—besides their beauty and durability—a concrete characteristic that makes them highly desirable. Gems combine high value with small size. They are easy to transport, easy to conceal and easily convertible to money. In times of stress, when the value of money has weakened, gems have often been the means of saving the substance of an estate. Land and houses may not survive a revolution, but a secret hoard of gems may outlast any political system.

With expanding populations and increasing prosperity in many countries of the world, the number of potential customers for gems—as for paintings and other works of art—is increasing far more rapidly than the available supply. Result: a vast annual trade, and rising prices. Unfortunately the buying public generally does not know much about gemstones.

It is not essential for the gem buyer to master truly technical information. But there are certain basic facts that can be easily assimilated and can reinforce him with a certain amount of knowledge even though he must still rely on the jeweller in technical matters. Certain essential characteristics, italicized below, must always be reviewed when the quality and value of a gemstone are considered.

The *hardness* of a gem is the resistance it gives to scratching and general wear. Mohs' Scale, a rough scale of hardness used for all minerals, is the convenient and generally accepted guide.

MOHS' SCALE OF HARDNESS		
Soft	1.	Talc
	2.	Gypsum
	3.	Calcite
	4.	Fluorite
	5.	Apatite
	6.	Feldspar
	7.	Quartz
	8.	Topaz
	9.	Corundum
Hard	10.	Diamond

PRECEDING PAGES: *Rough gem material must be expertly analyzed to determine how best it can be converted into marketable gems. Leading diamond-cutters like Bernard de Haan work only on diamond rough.*

All gemstones of any importance have a hardness above 6 in this scale. If less than 6 in hardness, they are not durable enough for use as gems. Any gem material that is not scratched by a sample of quartz, but is by topaz, would have a hardness of $7\frac{1}{2}$. For testing purposes, to help identify a gemstone by its hardness, the jeweller may use a set of pencil-like tools with hardness points made of minerals in the scale.

The *weight* of gemstones is usually given in carats, one carat equalling one-fifth of a gram. Converting this into more familiar units, there are about 140 carats in an ounce. Even knowing that the carat is an expression of weight and not size, gem buyers are always surprised to see that a 1-carat sapphire is considerably smaller than a 1-carat diamond. Sapphire is denser than diamond and a smaller stone can therefore weigh the same number of carats as a larger diamond.

Determining the *specific gravity* of a gem involves two measurements at the same time—size and weight. We are aware of a difference in weight when we compare iron and wood, yet it would not always be correct to say that iron weighs more than wood since a large piece of wood can weigh more than a small piece of iron. Only by comparing equal volumes of these materials can the extent of the weight difference be clear and unmistakable. Diamond is three and one-half times heavier than an equal volume of water—normally used as the standard of measurement—hence its specific gravity is 3.5.

The ability of mineral structures to bend a beam of light can easily be demonstrated (see page 113). The amount of bending or refraction—the mineral's *index of refraction*—depends entirely on the species of mineral involved, because it is an effect brought about by the mineral's structure. The degree of bending, measured by a refractometer, is useful in

THE
MIRROR
OF
STONES:
IN WHICH
The Nature, Generation, Properties, Virtues and various Species of more than 200 different Jewels, precious and rare Stones, are distinctly described.

Also certain and infallible Rules to know the Good from the Bad, how to prove their Genuineness, and to distinguish the Real from Counterfeits.

Extracted from the Works of *Aristotle, Pliny, Isidorus, Dionysius Alexandrinus, Albertus Magnus*, &c.

By *Camillus Leonardus*, M. D.

A Treatise of infinite Use, not only to Jewellers, Lapidaries, and Merchants who trade in them, but to the Nobility and Gentry, who purchase them either for Curiosity, Use, or Ornament.

Dedicated by the Author to CÆSAR BORGIA.

Now first Translated into *English*.

LONDON:
Printed for *J. Freeman* in *Fleet-street*, 1750.

TOP: *Woodcut of a lapidary shop from the medieval medical work* Hortus Sanitatis: *stones, like plants, were thought to have curative properties.* RIGHT: *A Renaissance classic, Leonardus' celebrated gem book appeared in English in 1750.*

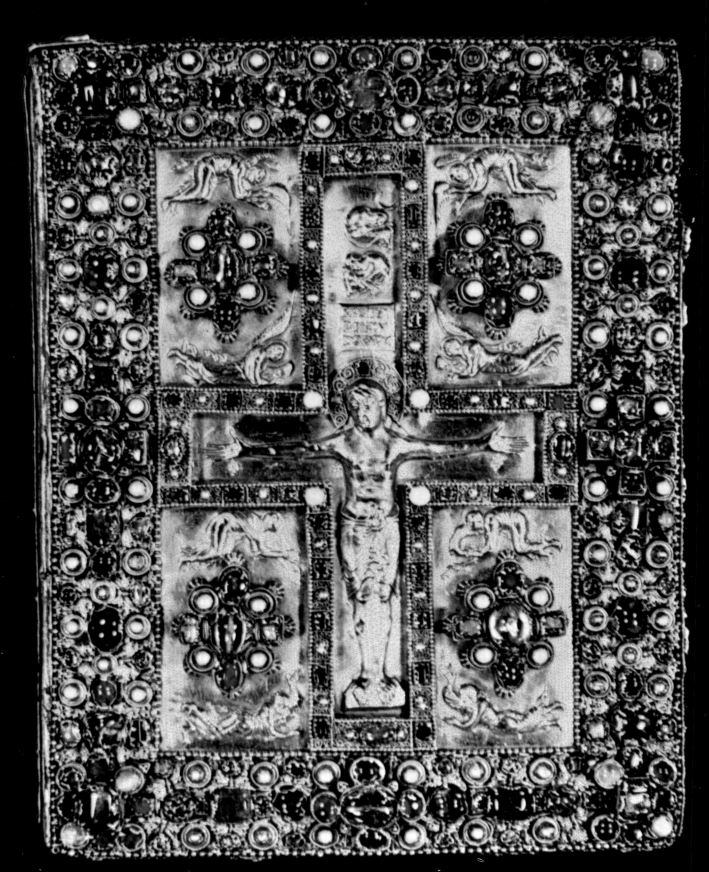

gem identification, since every gem has a different index of refraction.

When Newton split a beam of white light into its various wavelengths, or colours, he demonstrated the phenomenon of *dispersion*. The amount of refraction when a beam of light strikes a gemstone depends not only on the gem's structure, but also upon the wavelength of the light. The blue wavelength, for example, is bent more than the red. Since white light is composed of all wavelengths, each tends to be bent a different amount and the light is thus separated and sorted into a rainbow of colours. If the gemstone has strong dispersive ability, like diamond, it will cause a wider separation of the wavelengths and sparkle with coloured flashes whenever struck by a beam of white light.

Chatoyancy—from the French meaning cat's eye— refers to the single slit of light, like the contracted pupil of a cat's eye, reflected from certain rounded,

oval-cut gemstones. Closely related is the *asterism*, or starlike reflection, so well known in star rubies and star sapphires.

These effects are the results of inclusions— foreign bodies—found in some mineral crystals. Generally inclusions are considered undesirable in gemstones except as an aid in identification. However, some inclusions can greatly enhance the value and beauty of a gem by producing striking light effects.

Chatoyant mineral specimens contain inclusions of needle-shaped reflecting bodies of other species, such as rutile. These inclusions may even be hollow tubes. The combined reflection phenomenon— chatoyancy—is produced when the tubes or needles are orientated in parallel bundles. By cutting the gemstones as oval or circular domes, the gem cutter is able to reveal the resemblance to a cat's eye.

Asterism is an effect that can occur when there are

FACING: *Goldwork in the Celtic-German tradition on the ninth-century* Lindau Gospels; *light filters through jewels in raised mountings.* ABOVE: *Five-inch beryl crystals from Brazil, rare because they are still attached to albite feldspar.*

several sets of needle inclusions. In crystals of corundum—from which we get the gemstones ruby and sapphire—these sets arrange themselves in three different directions, 120 degrees apart, parallel to possible crystal faces. Instead of one band of light there are now three bands, intersecting at a common point. If the gem is properly cut the effect is a reflection of a bright star of light.

GEMSTONES: Armed with these brief explanations, we can now look more closely at individual gem species. The list is by no means exhaustive; it is compiled from personal preferences.

Diamond: Diamond, a naturally occurring variety of carbon, is not only 'a girl's best friend' but a jeweller's as well; it is the backbone of his trade. Through historical accident, its own fine characteristics, and effective promotion, it is now at least as highly treasured as it was in the past.

Its value and its rarity have always been legendary. An ancient story attributed to Aristotle says, 'No one except Alexander ever reached the place where the diamond is produced. This is a valley connected with the land of Hind. The glance cannot penetrate to its greatest depths and serpents are found there, the like of which no man hath seen, and upon which no man can gaze without dying. However, this power endures only as long as the serpents live, for when they die the power leaves them. In this place summer reigns for six months and winter for the same length of time. Now, Alexander ordered that an iron mirror should be brought and placed at the spot where the serpents dwelt. When the serpents approached, their glance fell upon their own image in the mirror and this caused their death. Hereupon, Alexander wished to bring out the diamonds from the valley, but no one was willing to undertake the descent. Alexander therefore sought council of the wise men, and they

told him to throw down a piece of flesh into the valley. This he did, the diamonds became attached to the flesh, and the birds of the air seized the flesh and bore it up out of the valley. Then Alexander ordered his people to pursue the birds and to pick up what fell from the flesh.'

The most fantastic diamond ever found weighed more than one and one-third pounds before it was cut. It was the Cullinan diamond, named for Sir Thomas Cullinan, chairman of the Transvaal Company at whose Premier Mine in South Africa the stone was found in 1905. Its shape suggested that this diamond was only part of an original crystal twice the size. It was bought by the government of Transvaal and given to King Edward VII of England in 1907. The great rough stone was then cut. From it came the Star of Africa, 530.2 carats, the largest cut diamond in the world, now set in the King's Royal Sceptre. In addition, the mass provided a 317.4-carat gem now set in the Imperial State Crown; two large stones of 94.45 and 63.65

of India. Legend, however, traces it back almost five thousand years.

Large diamonds like the Cullinan and the Kohinoor easily catch the public fancy. However, these stones are commercially insignificant compared to the wealth of smaller diamonds required for the gem trade and for industrial uses. Diamond mining that relies on predatory birds, as in Alexander's method, makes a romantic legend. It is questionable whether even Alexander ever really tried it. In any case, modern, practical methods for mining diamonds require earth-moving machinery and other processing equipment.

Even considering the more than three-ton total of gem and commercial diamond mined each year, diamond is a rare mineral. In the rich South African mines, it is found only as one part in fourteen million of the ore. This partially accounts for the nearly constant price of diamond, which is almost as high as that of the most expensive gems in the world, rubies and emeralds. And in spite of this price, the demand for diamond remains so great that operating producers and freelance prospectors are always out searching for more.

Historically, the earliest known diamonds came from river gravels in India and Borneo. The Golconda district in India was one of the first regular producers. India's other important diamond districts are one farther north, between the Mahanadi and Godavari Rivers, and one in central India, near the city of Panna. From ancient times until the discovery of diamonds in Brazil in 1728, these Indian mines were major producers. Now their yield is so small that profits barely justify costs.

Diamond mining in Brazil began when slaves looking for gold found some unusual, bright, shiny pebbles. When these were identified as diamonds, the rush began. Diamond mines opened

carats, which are set in the Queen's Crown, where the Kohinoor, another fabulous find, is also displayed; five stones weighing 18.85, 11.51, 8.8, 6.8, and 4.4 carats; ninety-six additional stones weighing 7.55 carats in all. Other sensational discoveries have occurred with some regularity for centuries.

The Kohinoor itself has a history lost in antiquity. This diamond, which weighs 108.9 carats in its present cut form, first officially appeared in 1304, when it came into the hands of the Mogul emperors

FACING: *Jewels like these garnet-set gold Etruscan earrings, c. 200 B.C. are valuable archaeological clues.*
ABOVE: *From the rare* Grundriss der Edelsteinkunde *(1887), a group of mineral crystals, precisely drawn.*

at the headwaters of Rio Jequitinhonha in Minas Gerais and of Rio Sao Francisco in Bahia. Here as in India, the diamonds were found in the washed river gravels. It was the custom that when a slave found a particularly fine stone, he was honoured, fitted out with new clothes, and given his freedom. One local legend tells of the woman slave who found the 254-carat Star of the South; she was not only freed, but pensioned for life.

The Brazilian diamond mining industry was eclipsed in the 1800s when rich diamond deposits were discovered in South Africa. Brazilian mining methods were primitive, accessible ore fields had been mined out, and the deposits could not produce as bountifully as those of Africa. There was a succession of diamond strikes on the African continent, starting with the great South African rush of 1866. And still the intensive search for diamonds continues.

A diamond is often called the 'king of gems'. Composed of almost pure carbon, it has to a high degree everything a mineral substance requires to be considered a gemstone—brilliance, beauty, durability, rarity, portability. It also happens to be fashionable—an added advantage. Because of its high index of refraction and high dispersion it produces brilliant flashes of bright coloured fire. Its durability is beyond question, since it is the hardest natural substance known. But hardness is merely a measure of resistance to wear and abrasion; this does not mean that diamonds are unbreakable. Breakage has to do with how well the material holds together. Diamond resists scratching because there is no natural substance hard enough to scratch it. Diamond is brittle, however, and it does have a cleavage—what the diamond-cutters call a grain. Under the shock of a sudden sharp blow a diamond may crack or chip or even cleave.

Diamond can be found in a wide range of colours. Most gem diamonds are water-clear and colourless. Some have a very pale, steely blue tint, which gives rise to the term blue-white. Colour—or lack of it—is important in the commercial grading of diamonds. Two diamonds, alike in their brilliance and the beauty and perfection of their cut, and totally lacking in inclusions, may differ a hundred per cent in price because of a subtle difference in colour. Really fine coloured diamonds, or 'fancies', are quite rare and bring premium prices. Yellow and brown colours are commonest, but if only pale in hue they actually reduce the value of the diamond. Greenish diamonds are sometimes found; rarest of all are red, pink and blue stones.

Considering the rarity of truly blue diamonds, it comes as something of a surprise to see the Hope diamond for the first time in its black felt-lined private vault at the Smithsonian Institution in Washington. It weighs 44.5 carats and is a deep, rich blue. People often compare its colour to sapphire, but it really is a dark, hard indigo blue, very much the same colour that you would expect to see on heated steel. The Hope diamond's fantastic history reads like a detective story with inexplicable

Two specimens still embedded in their matrix materials. FACING: *From Minas Gerais, Brazil, brilliant varicoloured tourmaline in lepidolite mica.* ABOVE: *Red crystal sections of ruby embedded in chromezoisite, from Tanzania.*

gaps. It was brought to France from India in 1642 by Tavernier, a noted gem explorer, who sold it to Louis XIV in 1668. After it was cut into a triangular-shaped gem weighing 67.13 carats, it made its way into the collection of French crown jewels. In 1792, during the French Revolution, it was seized and deposited with other treasures at the Garde-Meuble, from which it subsequently disappeared.

In 1830 a blue diamond, somewhat lopsided and weighing 44.5 carats, was sold to Sir Henry Thomas Hope in London. It seemed to be a cut-down piece of the French Blue. In 1874, when the Duke of Brunswick's gems were sold at auction in Geneva, a blue stone weighing between 6 and 7 carats was among them. In colour and quality it was astonishingly like the Hope diamond, and very likely was the missing tip of the French Blue.

The Hope itself made its way through several hands, leaving behind a legendary trail of misfortune and violence, until it was acquired in 1949 by Harry Winston, the New York jeweller, who later presented it to the Smithsonian Institution.

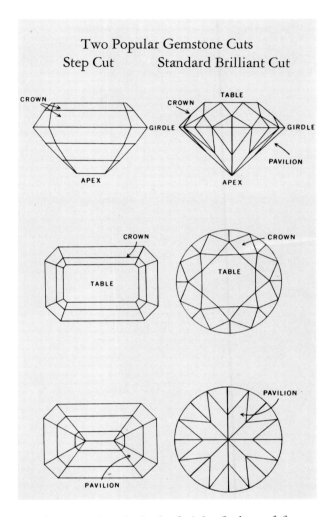

Two Popular Gemstone Cuts
Step Cut   Standard Brilliant Cut

With diamonds so rare, so intriguing, and so much in demand, it is inevitable that many attempts are made to find or manufacture less expensive substitutes. Some of the substitutes are sufficiently deceptive so that at a short distance it is impossible to tell them from the real thing.

The natural gemstones that most frequently substitute for diamonds are zircon, white sapphire and white spinel. Synthetic rutile and strontium titanate are also used, as is glass. White sapphire and white spinel can be detected by careful examina-

tion because they lack the bright flashes of fire so characteristic of diamond. Strontium titanate, on the other hand, has entirely too much flashing colour; it has four times the colour-dispersing ability of diamond and looks gaudy by contrast. Synthetic rutile is always a little yellowish, even though it has some of the necessary fire. Rhinestones, at one time cut from quartz but now cut from glass, have no fire, lack brilliance, and, since they are of soft material, often show innumerable scratches under magnification. This leaves zircon as the best

ABOVE: *A nineteenth-century artist's conception of a smokey quartz grotto in the Swiss Alps. Such caves led to the old notion that quartz was frozen water.* RIGHT: *Coloured stones are often step cut; the many-faceted brilliant cut best displays the fire of such stones as diamonds.*

substitute because it has clarity of colour, sufficient hardness and brilliant fire. The expert, however, can always identify zircon by examining the edges of the back facets, through the top of the stone, with a magnifier. They appear doubled because zircon is a double-refracting substance. Diamond is not.

The laboratory manufacture of synthetic diamonds has reached the point where a steady production of small commercial-grade stones is possible. Crystals of the size needed for gemstones, however, will have to be mined for some time to come.

Ruby and sapphire: The name ruby comes from the Latin word *ruber*, meaning red. Sapphire is also from the Latin and means blue. It is valid to consider them together because ruby and sapphire are the same mineral species—corundum—which is an oxide of aluminium (Al_2O_3). The red colour in ruby is due to traces of chromium replacing a bit of aluminium in solid solution. Sometimes the colouring is more rose than red. It varies through carmine to a dark purplish red called pigeon's blood. Stones this deep, rich red colouring come from Burma.

Rubies of good quality and colour are very rare; when such stones exceed a weight of 10 carats they are among the world's rarest gems. Fewer than half a dozen cut rubies have been recorded as important enough and large enough to carry individual names. Two of these are in the United States: the 100-carat De Long ruby at the American Museum of Natural History in New York, and the 138.7-carat Rosser Reeves ruby at the Smithsonian in Washington. These are possibly the largest and finest star rubies in the world.

Although sapphire means blue, any kind of corundum, except red, has come to be called sapphire. The finest sapphires come from Soomjam, in Kashmir, and are a soft, velvety, cornflower blue, which in the gem trade is called Kashmir blue. Corundum is colourless only when pure; traces of iron, chromium, titanium and other metals will produce blue, yellow and gold gems. Sometimes a single stone will show zones of two colours, such as yellow and blue. Unfortunately there has been a tendency to give a different name to each corundum colour variation, which confuses identification. Thus we may hear green sapphire called 'oriental emerald' and yellow sapphire called 'oriental topaz'. Also, because certain other gems, particularly spinel, strongly resemble ruby and sapphire, there is often further mis-identification. Neither the famous Black Prince's 'ruby' nor the Timur 'ruby' in the British crown jewels, for example, are really rubies. Both are red spinel. The very large 'ruby' that King Gustav III of Sweden gave to Catherine the Great of Russia is not a ruby either. It is a remarkably fine red gem of another mineral, tourmaline.

Corundum is usually opaque and poorly coloured, but being the next hardest natural substance to diamond, it is an important commercial abrasive.

FACING: *White-tipped amethyst from Guerrero, Mexico.* TOP: *From a recent Panamanian excavation, a thousand-year-old gold and emerald pendant.* ABOVE: *Gem tourmaline from Minas Gerais, Brazil.*

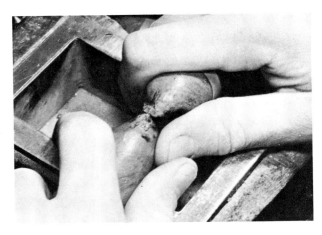

It is far more plentiful than diamond and can therefore be economically mined for industrial uses. The crystals are hexagonal in symmetry and, although very hard, are somewhat brittle. The specific gravity is about 4.0, which is among the highest for gemstones. This explains why a relatively small gemstone of ruby or sapphire may be several carats in weight.

Good-quality corundum—as indicated above, hard to come by—meets many of the criteria for gemstones. It occurs in transparent stones of several beautiful colours and its hardness enables it to take a high polish, which is extremely durable.

Near Mandalay, Myitkyina and Mogok, in Burma, ruby and sapphire mines have been in operation for centuries, but the most productive of them, near Mogok, are now being mined only sporadically. From the abundance of artifacts found there it is obvious that they were worked as early as the Stone Age. The stones themselves come mainly from gravels that were weathered out of the underlying metamorphic limestones of the region.

Almost all the other noteworthy occurrences of ruby and sapphire are in Asia. Ceylon has produced quantities of fine sapphire, second only to Burma, and primitive mining is still carried on in the gravels there. Thailand, from mines near Chantabun, has produced some dark rubies, though it is better known for the sapphires found near Battambang.

One of the interesting earmarks of a ruby or sapphire from an Asian mine is the way it is cut. Most stones are cut at the sites where they are found. Since the natives are paid for the weight of the finished stone, they usually cut with no regard for overall symmetry or beauty, but only to recover the maximum weight from the rough piece. As a result, most of these native-cut stones are peculiarly shaped, with deep bellies on the backs of the stones. The natives are very clever, too, at cutting parti-coloured blue and white stones in such a way that the blue becomes the bottom point. It thus suffuses the entire stone with colour.

The best green and orange-yellow sapphires come from Australia. The Anakie gemfield, two hundred miles from Rockhampton in central Queensland, has for over seventy years produced many of Australia's finest sapphires. Harry Kazanjian, the Los Angeles gem merchant, cut and polished the 733-carat Black Star of Queensland, the largest star sapphire in the world, which came from Anakie. He had four large stones from this field carved into the heads of the American Presidents Washington, Jefferson, Lincoln and Eisenhower, averaging two by two and a half inches in size.

Colourful superstitions, which have trailed the ruby and the sapphire from ancient times, have died hard—or not at all. Some Burmese are still convinced that a ruby, embedded in the flesh, will protect its wearer from bodily harm. Among the Sinhalese the star sapphire is regarded as protection against witchcraft. This may account for the comparative inactivity of witches in the United States, where two of the most magnificent star sapphires in the world are located. They are the 536-carat Star of India in the American Museum of Natural History in New York and the 330-carat Star of Asia in the Smithsonian Institution.

Emerald: More than thirty-five hundred years ago, about one hundred miles north of Aswan, Egypt, on the slopes of Jebel Sikait and Jebel Zubara,

feverish mining activity was producing emeralds for most of the jewellery of the ancient world. The so-called Cleopatra's Emerald Mines are still visible —hundreds of mine shafts, some going down eight hundred feet. In them have been found mining tools dating back to 1650 B.C. These mines are dead; but the emerald, after all this time, is still extremely valuable and eagerly sought.

Emerald is a green variety of the mineral species beryl, which exists in many colours and in considerable quantity. The special value of emerald lies in its rich green colour, which is very rare. Emerald, and all other beryl, is chemically a beryllium aluminium silicate ($Be_3Al_2Si_6O_{18}$). The various colours are produced by traces of other elements in solid solution. For emerald, the colouring agent is a

Three steps in the delicate art of diamond cleaving.
FACING, TOP: *Cleavage is analyzed and marked;* BELOW: *Rough crystal is scored.* ABOVE: *Cleaving. The result can be two or more pieces of fine gem rough—or comparative disaster.*

These vary in quality from very poor to very good, but even the best coloured stones, such as those from Russia, never attain the size of those from Colombia. The United States has produced large emerald crystals of good colour in North Carolina, but because of fractures and inclusions their clarity leaves much to be desired.

By far the bulk of all beryl mined is valuable as a source of the metal beryllium, but useless as a source of gemstones. However, it is not unusual to find crystals that can be cut into superb gems weighing over a thousand carats. One crystal from near the village of Marambaya in Minas Gerais, Brazil, weighed over two hundred pounds. It was laboriously brought to the coast and shipped to the great gem-cutting centre at Idar-Oberstein in Germany, where it was cut into 200,000 carats of many small stones. This crystal was an aquamarine, the blue-green variety of beryl caused by solid solution of a small amount of iron. The very deep-blue gemstone aquamarine, seen in jewellery shops, is always produced artificially, by heating greenish or certain yellow-brown crystals to a temperature of 800 degrees Fahrenheit. The resultant colour change is permanent.

Aquamarine and the other non-emerald beryls are found in pegmatites. Their large, well-formed crystals are typical of pegmatite minerals. Emeralds, found in metamorphic rocks, have crystallized under much more difficult circumstances and therefore are not as well formed.

Best known among modern sites for gem beryl are the prospects near Teofilo Otoni and Minas Novas in the state of Minas Gerais, Brazil. Good stones have also come from the Malagasy Republic (or Madagascar), from the Ural Mountains and Transbaikalia in Russia, and from the United States, chiefly from New England and California.

small amount of chromium. Emerald, with a hardness of 8, is one of the more durable gemstones. It crystallizes in the hexagonal system, often occurring in very well-formed euhedral crystals. Its specific gravity is 2.7, not nearly as high as sapphire, so that stones of large size can be worn without discomfort.

The finest emeralds ever mined have come from mines in Colombia—at Chivor, Muso, and elsewhere—since long before the time of the Spanish conquest. There they occur in veins of calcite in a rather dark-coloured, carbon-bearing limestone.

There are many other occurrences of emerald in the world, in Australia, Austria, Brazil, Russia, South Africa, Southern Rhodesia, India and Norway.

FACING: *Early Brazilian diamond mining.* THIS PAGE, TOP: *Waterkloof, South Africa, 1852—one of many battles lost by local Kaffirs to the British in what later became diamond country.* ABOVE: *Kimberley diamond market, 1888.*

There have been numerous and fairly successful attempts to find substitutes and imitations for natural beryl gemstones. The most successful synthetic emerald is the Chatham emerald, named for its originator. This is a beautiful gemstone with the appearance of high-quality emerald. Usually the stones are small and, although considerably less costly than natural stones of the same quality, are still fairly expensive.

A synthetic blue spinel is easily made and can be successfully substituted for a rich blue aquamarine. The 'soudé' emerald is also a successful substitute, although it is easy to detect by sight. It consists of a sandwich of some green material, often a copper compound, between two pieces of cut and polished quartz which form the top and bottom of the composite stone. Sometimes the backs of pale or off-coloured emeralds and other beryls are painted with the colour needed to strengthen them. There are even glass imitations for beryl that are quite successful, with the glass deliberately cracked and reheated to give the appearance of stones with natural flaws.

Pearl: Pearl is not strictly a mineral. It is derived from the excretions of a live animal and 10 to 14 per cent of its mass is an organic substance called conchiolin. However, the greater part of the mass of a pearl consists of between 82 and 86 per cent calcium carbonate—the mineral species aragonite—and it is included here because it has always held place among the more important and desirable gemstones.

Many kinds of shelled animal species can and do form pearls. However, most of these pearls are so grossly inferior in colour, lustre and shape as to be useless for gem purposes. Only two species of molluscs seem able to produce pearls of value. In salt water the species Margaritifera, an oyster not related to our edible oyster, produces pearls in the best quantity and quality. The species Unio, a kind of mussel, is the freshwater source of good pearls.

Pearl formation is due to particles of sand, to tiny parasitic worms that bore through the shell, or to other irritating bits of material that stimulate the animal into trying to relieve itself of the annoyance. Sometimes the irritant is pushed against the inside of the shell and plastered over with layers of pearl to form flat or button-shaped 'blister' pearls. At other times the irritant forces its way into the tissues of the animal. In this case it is apt to become covered equally on all sides with pearl to form a sphere. If for some reason the covering is not spherical, but is partially or extremely misshaped, the result is known as a baroque pearl.

The material that the mollusc produces to cover the irritant, as well as to line the entire inside of its shell, is called nacre. This consists of alternate layers of aragonite, deposited in many tiny overlapping plates, interbedded with layers of conchiolin. These tiny plates create the pearl's beautiful lustre by causing light interference. They also help in identification, for because of them a real pearl will feel a little rough when it is scraped over the teeth.

The value of pearls depends on their shape, colour, sheen, size and origin. Colour in pearls is an elusive quality that can usually be identified only by experts. Pure white or white with a faint tinge of pink or yellow are considered the most

desirable colours. From time to time, usable pearls are found in fancy colours: yellow, bronze, pink, green, blue, black. These strongly coloured pearls, however, rarely have a good lustre, or 'orient'. Without this bright, lustrous sheen, a pearl drops considerably in value.

When the Japanese perfected a process that stimulated the oyster to place a thick layer of nacre over an artificially introduced irritant, the cultured-pearl industry was born. Experimentation established that the best starting material is a bead of mother-of-pearl—the nacreous material with which the mollusc lines its shell—cut from the shell of the pearl mussel of the Mississippi Valley. These beads, cut in various sizes, are inserted skilfully into the live oyster, and it is returned to the sea to continue growing. Treated oysters are kept in large growing rafts. From each of these rafts are suspended about sixty cages holding a total of three thousand oysters at a depth of seven to ten feet. They are inspected several times a year; after three to seven years the beads will have been coated with a layer of nacre

FACING: *The great Hope diamond, 44.5 carats—surprisingly, steel blue.* ABOVE: *The Hope's story is still mysterious. Louis XIV's stolen French Blue reappeared in 1830 as the smaller Hope, in 1874 as a 6-carat piece in the Duke of Brunswick's collection.*

almost one-sixteenth of an inch thick, and they are brought up and harvested.

Cultivated or natural, the pearl ranks as a major gemstone. If the diamond is called the king of gems, pearl is queen.

Jade: The name jade refers not to one mineral species but to two—jadeite and nephrite—which have practically no relationship to each other except for a similarity of appearance. Jadeite is a sodium aluminium silicate occurring in white, emerald-green, and other colours, and is one of a group of rock-forming minerals called pyroxenes. Nephrite is a calcium magnesium iron silicate occurring in colours ranging from white to spinach-green to black, and is one of a different group of rock-forming minerals

called amphiboles. Both jadeite and nephrite are only about $6\frac{1}{2}$ in hardness, but are so tough that they will stand amazing amounts of abuse.

The ancient Chinese worked *yu* as early as 1000 B.C. Most of this jade apparently came from Khotan and Yarkand in Turkestan. Marco Polo visited the area and described the mining methods. Under the sharp eye of a supervisor on shore the searcher would wade out into the stream, feeling for the jade boulders with his feet. When he touched what felt like one, he would work it to shore. This material was all nephrite. It was not until about the eighteenth century that the fine-quality jadeite found in Burma reached China, beginning a classic period in which beautiful objects were carved from superb, imperial green jade. This kind of jade is still so highly prized that a one-inch piece may be worth as much as two thousand pounds. The art of jade carving was subsidized by Chinese royalty during the Ching Dynasty (1644-1912), so that it reached its peak in relatively modern times.

Opal: In the gem market the supply of hard-to-find precious opal fluctuates too much for it to establish a steady price structure. When good-quality opal is available, its life is unpredictable, for the gem, being quite brittle, may crack if subjected to heat or a sharp blow. Further, it is not very hard, and surface scratches from constant wear can dull its sheen. Nevertheless, the beauty of opal is so exquisite that in recent years it has become a popular gemstone.

Opal is a compound of the elements silicon and oxygen, somewhat like quartz but different in its structure. The gem trade recognizes four kinds of precious opal: white, black, fire and water. White opal has a pale or white body colour as a background for its fire. Black opal may be dark blue, dark green, dark grey or black in body colour. Fire opal and water opal are either transparent or translucent. Fire

opal has a yellowish, orange or orange-red colour and may or may not exhibit fire, whereas water opal is colourless but has fire. Not all opal has beautiful colour and fire; if it does not, it is called common opal.

All gem opal comes from a sedimentary environment, but there are surprisingly few good sources. Gem opal has been known for centuries, as indicated by its name, which comes from the Sanskrit word *upala*, which means precious stone.

In 1889, white opal was found at White Cliffs, Australia; this was soon followed by other Australian discoveries. But opal did not have much of a commercial market until later in the nineteenth century, when black opal was found at Lightning Ridge,

Australia. Recently, there has been a growing interest in the fire opal from Mexico. The Mexican deposits have been worked since 1835 and have produced superb gems, some of them over 50 carats.

To help protect opal from damage, it is mounted most frequently in pins and brooches, rather than in rings or bracelets, and is usually given some kind of metal backing.

The expensive, highly desirable gemstones that have been discussed are familiar—at least in name—to almost everyone. But there are many other gems that, for a variety of reasons, have remained less well known. Some of these follow, listed here alphabetically, rather than in the order of their popularity, value or beauty.

FACING: *Mediterranean coral diving—no longer profitable.*
ABOVE: *An exploratory adit—a horizontal tunnel—into the ruby-bearing gravels of Burma, at a time when the British were trying to revive flagging industry there.*

Amber: This material is actually sap that flowed, in times past, from various species of trees, and became fossilized. The world-famed Baltic variety, for example, came from a species of pine tree. Amber is therefore of organic origin and not really a mineral, but like the pearl holds a valid place in any discussion of gem materials. The colour varies through shades of yellow, brown and red. Amber is well simulated by so many modern gums and artificial resins that it is extremely difficult to identify with certainty. Some artificial amber is even manufactured with insects embedded, as often occurs in natural amber. In determining the genuineness of such a specimen, the sceptic's only hope would be to identify the insect to see if it is truly an ancient species.

Benitoite: Discovered in 1907, benitoite has been included in the company of gemstones only in modern times. It is a chemically complex barium titanium silicate found only in San Benito County, California, which gives it its name. A hardness of $6\frac{1}{2}$, strong fire, high brilliance, and a lovely sapphire-blue colour make it an excellent gemstone. Unfortu-

92

Examples of pegmatite minerals. ABOVE: *Gem morganite from San Diego County, California.* FACING: *Two gems from Minas Gerais, Brazil, step-cut for best display of their colour.* ABOVE: *3273-carat topaz.* BELOW: *2054-carat beryl.*

nately, it is so rare that it is available only as an expensive collector's item. The largest existing cut stone, now in the Smithsonian Institution, weighs 7.6 carats.

Chrysoberyl: Ordinary chrysoberyl has a hardness of $8\frac{1}{2}$ and occurs in a colour range from yellow through grass-green to brown. It is a beryllium aluminium oxide occurring in good-quality natural crystals, often in attractive cyclic twin crystals.

In 1830 a variety of chrysoberyl was found in an emerald mine near Takovaja, in the Ural Mountains of Russia, that was green in daylight and red in artificial light. Since Tsar Alexander II had just come of age, the gem material was named alexandrite in his honour. Later, in the gem gravels of Ceylon, larger and finer alexandrites were found. The largest of these cut stones recorded, 66 carats, is now in the Smithsonian Institution; another large stone, 43 carats, is in the British Museum (Natural History).

Alexandrite is so popular that it is as expensive as the traditionally most precious gems—diamond, ruby, sapphire and émerald. Partially successful attempts have been made to imitate it with a colour-changing variety of synthetic sapphire.

Another important variety of chrysoberyl has an extremely strong and sharp chatoyancy, which produces exquisite cat's-eyes when the stones are cut in round-topped cabochon form. The colours of these vary from a light honey-yellow-green to brown to almost black, but each colour shows the strong bright line necessary for a good cat's-eye.

Coral: Like amber and pearl, coral is not actually a mineral, since it has been deposited in the sea by coral animals. However, it has always been considered a gem material for beads, jewellery and carvings, and does consist of the mineral calcite. Most coral is white, but there are colour variations from pale pink to a deep oxblood red. Coral fisheries are limited to very definite areas, because coral animals require a water temperature between 55 and 60 degrees Fahrenheit. Regular coral fishing goes on

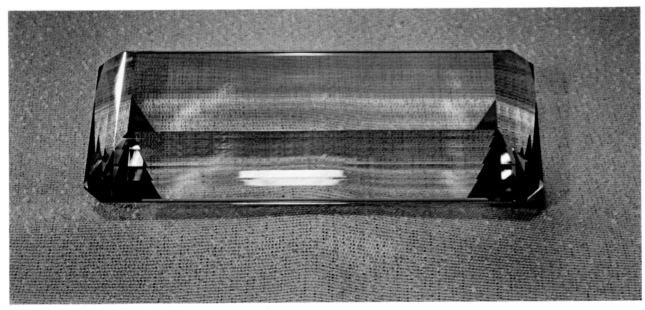

in Japanese, Malay and Mediterranean waters, with the finest red coral coming from Algeria and Tunisia.

Euclase: Although it has a hardness of $7\frac{1}{2}$, this beryllium silicate has a very strong tendency to cleave, which detracts considerably from its value as a gemstone. It is difficult to cut and cannot tolerate hard wear. Nevertheless, it is rare and beautiful enough to remain in demand, especially among gem collectors. The colour ranges from colourless to pale aquamarine to greenish-blue. Most of the gem-quality material is found near Ouro Prêto, north of Rio de Janeiro, Brazil.

Feldspar: The feldspars are actually a family of several closely related minerals, most of which are usable as gem material in one form or another. The most important gemstones among them are varieties of the feldspar species orthoclase. The best known of these is moonstone, which has a beautiful schiller (see page 126) aptly described by the name of the gem. Since the schiller is caused by

the interference of light, it is best revealed if the cabochon is cut parallel to the feldspar layers that cause the interference.

Another orthoclase variety is a clear, pale yellow to golden-yellow stone that comes only from the pegmatites at Itrongay, Madagascar. The colour in this variety, used as gems, is caused by a small amount of iron in solid solution.

Garnet: As is the case with feldspars, garnets also comprise a family of mineral species with at least six important members. They are pyrope, almandine, spessartine, uvarovite, grossular and andradite. All of these except uvarovite, the chromium-bearing garnet, furnish fine gemstones of various colours. All are hard enough—about 7—and all have a refractive index high enough to make them brilliant. If they have a major drawback it is that their colour is sometimes so dense that they look opaque. Ruby-red pyrope, used in antique Bohemian garnet jewellery, is perhaps most popular,

94

Chinese jade cutters work with extraordinary skill.
LEFT: *Jade is sliced with a lapidary saw.* ABOVE: *Grinding wheel is used to carve out delicate features.*
FACING: *The goddess K'wan-yin, an eleven-inch figurine of Burma jadeite.*

but the rich orange-red spessartine found in Virginia and elsewhere is more striking because of its unusual colour. A bright emerald-green andradite found in Russia and Italy is called demantoid and, when the quality is good, sells at prices rivalling diamond, sapphire and emerald.

Peridot: Peridot—a gem name for the common mineral species olivine—is easily simulated in glass and has little brilliance or fire. Its lack of lustre and peculiar rich bottle-green to brown colour give it a certain subtle charm, but a strong tendency to cleave and a hardness of only $6\frac{1}{2}$ have prevented it from achieving great popularity. Nevertheless, it persists as an offbeat sort of gemstone, not too expensive and generally available. Crystals suitable for gem-cutting come from Egypt, Burma and Arizona.

Quartz: Quartz, although a single mineral species, has several gem varieties whose names derive from differences in colour. It is common enough so that prices of the finished gems are within reach of nearly everyone. A hardness of 7 is sufficient for durability. Quartz, which is silicon dioxide, is the commonest mineral species in the crust of the earth. Many of its gem varieties, such as agate, onyx, tigereye, jasper, bloodstone, prase, are opaque varieties used for cabochons, carvings and art objects. The most popular faceted varieties are amethyst, citrine and rock crystal. Citrine varies in colour from a smoky brown to a golden yellow and often imitates the colour of topaz, so that it is sold as Brazilian topaz. Much of the faceted quartz now on the market comes from Brazil.

Sphene: This calcium titanium silicate is one of

the rarer minerals that produces good gemstones. It has a barely suitable hardness of $5\frac{1}{2}$, is brittle, and has some tendency to cleave. The beauty of the gems, however, makes these drawbacks seem irrelevant. They are yellowish, brownish or greenish, very brilliant and transparent, and because of high dispersion have a fire greater than that of diamond. The best faceting-quality sphene comes from Switzerland, the Austrian Tyrol and Baja California.

Topaz: There is a popular belief that all topaz is yellow and all yellow stones are topaz. Neither of these notions is true. Topaz, an aluminium silicate containing fluorine, occurs in fine gem crystals of several colours. The commonest is colourless, but light blue shades resembling aquamarine are also readily available. In addition there are light green, light rose and deep red varieties. Certain yellow Brazilian topaz turns bright pink when heated, and stones of this colour are prepared by natives on a somewhat hit-or-miss basis. They never know in advance whether the material they are heating will react with the desired colour change. The most highly prized stones are the colour of rich sherry or muscatel wine. Because it occurs in sufficient quantities, prices for topaz gemstones have remained moderate.

Topaz has a good lustre and its gems are brilliant and hard—8 on the scale—but it has a discouraging tendency to cleave easily, so that care must be taken in handling it.

Tourmaline: Part of the popularity of tourmaline is based on its very wide colour range. It is found in crystals of pink, magenta, green, blue, yellow, brown, black, and many subtle intermediate shades. Pink and emerald green are particularly popular. It is even possible to find crystals exhibiting two or more colours. Tourmaline is a complex borosilicate with a hardness of just over 7 and little tendency to

cleave; it is durable and useful as a gem in most kinds of mountings. Even though its refractive index is too low for brilliance and its dispersion is too low to produce fire, its beautiful colours make it very desirable. As with topaz, quartz and certain other gems, it occurs in sufficient quantity to keep prices moderate.

Zircon: The high refractive index of zircon, a zirconium silicate, gives it brilliance, and its high dispersion gives it fire enough to substitute for diamond. Its colours range from blue and golden brown to no colour at all. Zircon has a hardness of over 7, but its edges wear and chip very easily with careless handling. It is readily found in parts of Thailand, Viet Nam, Laos, and also in the rich gem-laden gravels of Burma and Ceylon.

THE LAPIDARY ARTS: Before a gem material becomes a gem, it must pass through the hands of the lapidary—the man who uses many techniques to cut each specimen so that it achieves its most beautiful, or most salable, potential.

Very little was done to take the fullest, most carefully calculated advantage of gem rough until the seventeenth century. Before that time the general procedure was to grind and polish existing crystal faces to increase their transparency and reflecting ability. It was also customary among ancient lapidaries to round off the tops of some of the more colourful, opaque gem fragments. We still do something of this sort today when we tumble pieces of gem material with abrasives until they assume a high polish but retain their free form.

The opaque gem materials, such as turquoise and jade, are still characteristically cut by being rounded off to produce a relatively large, dome-shaped surface with few conflicting reflections, so that the colour or pattern is well displayed. Such stones are called cabochons—a word derived from

Three exceptional gems, diversely treated. FACING, TOP: *A great blue heron of sodalite, cut in Germany.* FACING, LEFT: *330-carat Burmese Star of Asia, the second-largest blue star sapphire known.* FACING, RIGHT: *880-carat faceted kunzite from Brazil.*

the Old French *cabo*, meaning head—referring to their rounded tops. There are simple cabochons with dome-shaped tops and flat bottoms, double cabochons with domes at top and bottom, and hollow cabochons with domed top and scooped-out or concave bottoms. Very deep-coloured garnets are traditionally cut as hollow cabochons to reduce their thickness and thus allow enough light through to show the colour. There is also the tallow-topped cabochon, with a dome so low that it is shaped like a drop of wax that has fallen on a cold surface.

Aside from use with opaque materials, the cabochon is also used for cutting cat's-eye and star stones. A bead is a circular double cabochon with both domes equally developed. Very large beads, or spheres, are often mounted or displayed as art objects. The most extraordinary sphere in existence is a flawless one at the Smithsonian Institution cut from a thousand-pound fragment of Burmese quartz. It weighs one hundred and six and three-quarter pounds.

The finest and most expensive gem materials are always cut as faceted stones. These are transparent gems with a series of carefully placed flat, reflecting faces. In general, to describe a faceting cut, the stone is divided into sections. The flat top is called the table, and the facets appearing above the middle of the stone are in the part called the crown. The middle is called the girdle. Below the girdle the facets lie in the pavilion. If there is a tiny flat face at the bottom of the gem, opposite the table, it is called a culet.

The exact angles for placement of these facets are different for every kind of gem material. For different purposes there have also been developed a number of standard patterns of faces. The most frequently used symmetrical cuts are the standard brilliant cut and the step cut. Modern jewellery design often calls for the baguette, cut-corner triangle, kite, epaulette, keystone, marquise and many more. Frequently the kind of setting to be used determines the size and cut of the gem.

Cutting requires three basic operations: sawing, grinding and polishing. Sawing is usually done with a rapidly turning disc of iron or bronze edged with diamond. The stone is carefully pressed against this edge and the diamond saw scratches its way through. Oil or water is used to bathe the saw and keep it from getting too hot. The second operation is grinding. Again, this makes use of a rapidly rotating horizontal disc—or 'lap'—charged on its surface with a water paste of an abrasive like corundum or powdered diamond. The last operation, using the same kind of disc, is polishing. This time the lap is charged with an extremely fine abrasive such as jewellers' rouge, or tin oxide, or finely powdered pumice.

Polishing is not merely the removal of surface scratches. The surface being polished often reaches the melting point, causing the mineral to flow and deposit a very thin glaze over its own surface, which obliterates scratches and other imperfections.

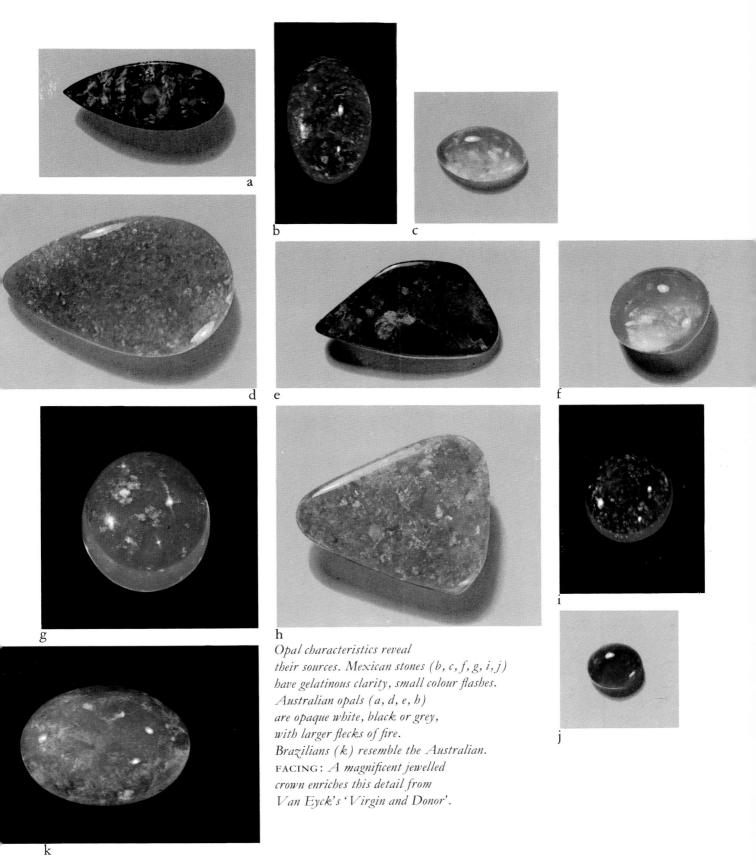

a

b c

d e f

g h i

j

Opal characteristics reveal
their sources. Mexican stones (b, c, f, g, i, j)
have gelatinous clarity, small colour flashes.
Australian opals (a, d, e, h)
are opaque white, black or grey,
with larger flecks of fire.
Brazilians (k) resemble the Australian.
FACING: *A magnificent jewelled*
crown enriches this detail from
Van Eyck's 'Virgin and Donor'.

k

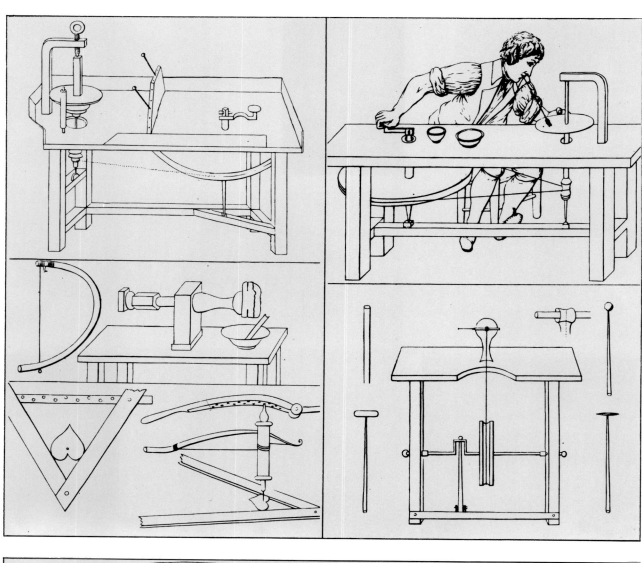

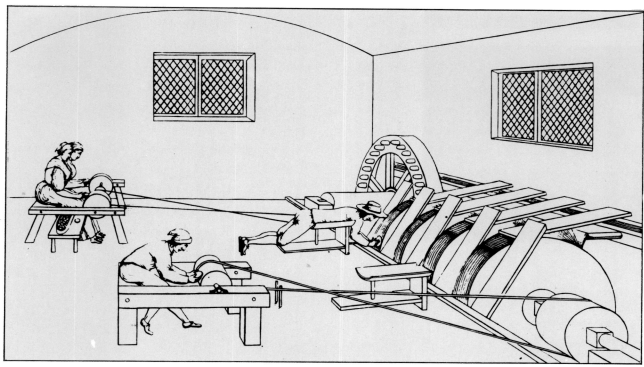

Perhaps the greatest form of artistic expression in working gem materials is carving. Everyone familiar with the jade carving of the Ching Dynasty in China appreciates not only the skill and patience of the carvers, but also the extraordinary characteristics of jade—toughness and tenacity—which permit this kind of carving.

Other cultures, at other times, have specialized in the carving techniques known as intaglio and cameo. Intaglio carving is incised into the gem and is usually intended for seals. Cameo carving is raised above the surface rather than incised. Both of these are often done in gem materials that contain layers of various colours, so that the raised or incised carvings are of different colour from the background. Agate and shell are used most often.

Anyone who cuts, polishes, carves or otherwise forms gem materials into objects of beauty considers himself—and is properly named—a lapidary. Because of the importance of diamond, however, anyone who cuts diamonds considers himself, and is known in the trade, as a diamond-cutter.

GEM IMITATIONS AND SUBSTITUTES: The number of natural stones appears really insignificant when compared with the number of gem imitations and substitutes that are being marketed. Imitation, assembled, reconstructed, and synthetic gems appear in great quantities, and usually without any attempt at deception or fraud. Unfortunately, even when not used intentionally for fraud, they have created confusion and apprehension in the minds of the gem-buying public.

Synthetics: All of the major gem minerals have been duplicated in the laboratory. To be commercially useful, laboratory methods must produce material of sufficient quality and size to be cut, and also to compete in price with gems dug from the earth. Good diamond, for example, is now

manufactured in quantity, but cannot yet be made in pieces sufficiently large to be cut into gems.

The production of synthetic ruby and sapphire for gems has been infinitely more successful. Cleverly synthesized rubies appeared in the early 1880s, thought to be made by heating fragments of natural ruby, but these 'Geneva rubies' were almost certainly not made entirely from fragments of natural ruby. Verneuil's chalumeau made the method obsolete. Sapphires are also easily made by Verneuil's method, substituting titanium and iron for the chromium oxide that is used in the ruby-making process. In the Verneuil method powdered aluminium oxide, containing the chromium, iron or titanium colouring agents, is sieved down through the flame of a vertical blowtorch furnace. As it passes through the flame, it forms a single crystal boule, or tapered cylinder of the synthetic gem material. In a few hours a boule of several hundred carats can be formed.

In recent years a method has even been found to make star rubies and star sapphires. The boules are prepared by adding between 0.1 and 0.3 per cent

FACING, TOP: *Grinding and engraving tools, from an eighteenth-century lapidary shop.* LEFT: *At Idar-Oberstein, Germany, a lapidary centre since Roman times, the Old Mill is still in use.* ABOVE: *The ancient crown of Charlemagne—part of the Imperial Treasure displayed in Vienna's Palace Museum.*

of titanium oxide to the feed materials in the Verneuil furnace. After forming, the boules are heated to just over 2000 degrees Fahrenheit, at which time the added titanium oxide forms thousands of tiny needle crystals that line themselves up in the direction of the sapphire or ruby structure. Actually, the stars look so much better than most natural star stones that they are very easily detected.

Spinel is another gemstone successfully synthesized. Natural spinel, except for its red variety, has never been popular. Because of this, most synthetic spinel is manufactured to simulate gemstones of other species. It can very easily masquerade as fine-coloured aquamarine, blue zircon, green tourmaline, pale green emerald, and can approximate a greenish-yellow variety of chrysoberyl. There is even a synthetic spinel that has the proper white, milky appearance of moonstone. All of this spinel is made by the Verneuil method, substituting magnesium oxide for part of the aluminium oxide.

There are two other gemstones made by the Verneuil method that are good enough to market but that do not appear in nature as gem material. In 1948, synthetic rutile was offered for sale under the name Titania. This is an appropriate name because, like natural rutile, the material is titanium oxide. The feed material is pure titanium oxide. It is passed through a Verneuil furnace that has been modified by the addition of an extra jet of oxygen. These boules come out black and are

later heated in a stream of oxygen, at which time they change from black to deep blue to light blue to green to pale yellow. Heating can be stopped at any point to retain a particular colour. The gemstones cut from these boules have a hardness of 6 to 6½, show very strong double refraction, and dispersion six times that of diamond. The flashes of fire caused by this extreme dispersion are far beyond anything diamond can approach.

The second material is strontium titanate, which is not found in nature at all, but has been sold since 1955 under the name Fabulite, as well as under other names, which suggest that it looks like diamond. This synthetic gemstone is bright and colourless and has a dispersion four times that of diamond, but is very easily scratched by a file.

Some other gems, like quartz, are made by a hydrothermal method. Quartz itself can be a gemstone, but synthetic quartz does not compete in price with natural quartz for this purpose. Synthetic emerald competes easily with natural emerald. There have been various successful attempts to make emerald, but the first stones to be successfully marketed were prepared by Carroll F. Chatham of San Francisco, California. He has been producing excellent emeralds of good colour, by a process he says is hydrothermal, using crushed beryl as feed material and perhaps a small natural crystal as a seed. Details of Mr. Chatham's method have never been divulged.

There is a newer process developed in the Austrian Tyrol that has fascinating possibilities. The process is again hydrothermal, using crushed beryl as feed material. In this case, however, the seed is a faceted stone of pale green beryl over which an emerald coating is grown. When completed, the overgrown facets need only be repolished and the stone is ready for sale.

Imitations: The commonest imitator of gemstones is glass, despite the fact that it is not always satisfactory. One variety of glass that is produced by melting natural beryl has none of the characteristics of crystalline beryl, but can be cut into surprisingly good gems. Normally, however, a common lead glass is used with various colouring materials, such as chromic oxide for green, cobalt oxide for blue, and didymium oxide for pink. The brilliance of a cut stone is sometimes enhanced by coating the back with a mercury amalgam, such as is used on mirrors, or by putting a layer of metal foil behind it in a closed jewellery mounting. Various kinds of plastics are also used with a similar effect.

Composites: For centuries it has been common practice to build up gemstones by fusing or cementing a shaped piece of natural gemstone to another piece, or pieces, of inferior or artificial material. If this is done with care, and the joints are artfully concealed in jewellery mountings, the composite stones can be very deceptive. The commonest kind is a doublet in which the top is a thin piece of garnet and the bottom is a coloured glass. The garnet, fastened to the glass by heating, furnishes a hard protective cover. Because this

FACING: *The finest known tourmaline from San Diego County, California—an extraordinary fourteen-inch specimen still on matrix.* ABOVE: *Sardonyx cabochons from Serra do Mar, Brazil—often simulated by dyed agate.*

cover is so thin, its colour is weak, so that the whole stone takes on the hue of the glass underneath.

Other doublets have been made in which both top and bottom are genuine, but are separate pieces glued together so that a larger gem can be made from two smaller or badly shaped pieces. Still others have the top made of the authentic gemstone other than garnet but the bottom of glass. A glass cabochon with a backing of iridescent abalone shell looks like a passable opal. Thin slabs of precious Australian opal are often fastened to a backing of the matrix rock from which the opal comes. Even triplets are sometimes made, in which the coloured material is sandwiched between a top and bottom layer of natural sapphire or natural quartz.

Artificial colouring: There are several methods used to intensify or change the colouring or apparent colouring of a gemstone. One direct method used for diamonds is to treat them in an atomic reactor, which converts off-colours into more beautiful hues of tan, yellow and green. The colour of the best golden-brown, sky blue and colourless zircon offered by jewellers has been induced in the stones by heating reddish-brown zircon pebbles under controlled conditions to nearly 1800 degrees Fahrenheit.

There are also indirect methods that add something to the stone that was not there originally. A coloured foil backing behind a stone of poor colour, in a closed mounting, cannot be seen and will change the gem's colour. It is even possible to paint or dye the back facets. A violet dye on pavilion facets of a yellow diamond will make it look almost colourless. It is not unusual to find colourless Mexican opal painted black on the back of the cabochon or fastened to a piece of black glass, which gives it a dark background to show off its internal fire.

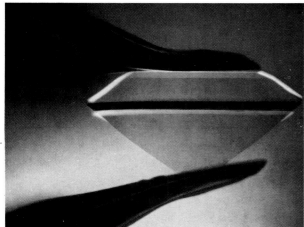

Turquoise impregnated with wax or plastic to deepen its colour, pale grey jade that is dyed green to look like imperial jade, and agate dyed red or blue or black or almost any colour—these are only a few examples of colour alteration in gems.

Faceted or carved, colourless or of brilliant hues, mounted or not, gems are universally admired.

FACING: *Dazzling light patterns refracted and reflected by a 2680-carat Brazilian topaz, the size of half a grapefruit.* THIS PAGE, TOP: *Emerald, a variety of beryl.* ABOVE: *An unset triplet clearly reveals its composition. Well set, it will appear to be a solid gem.*

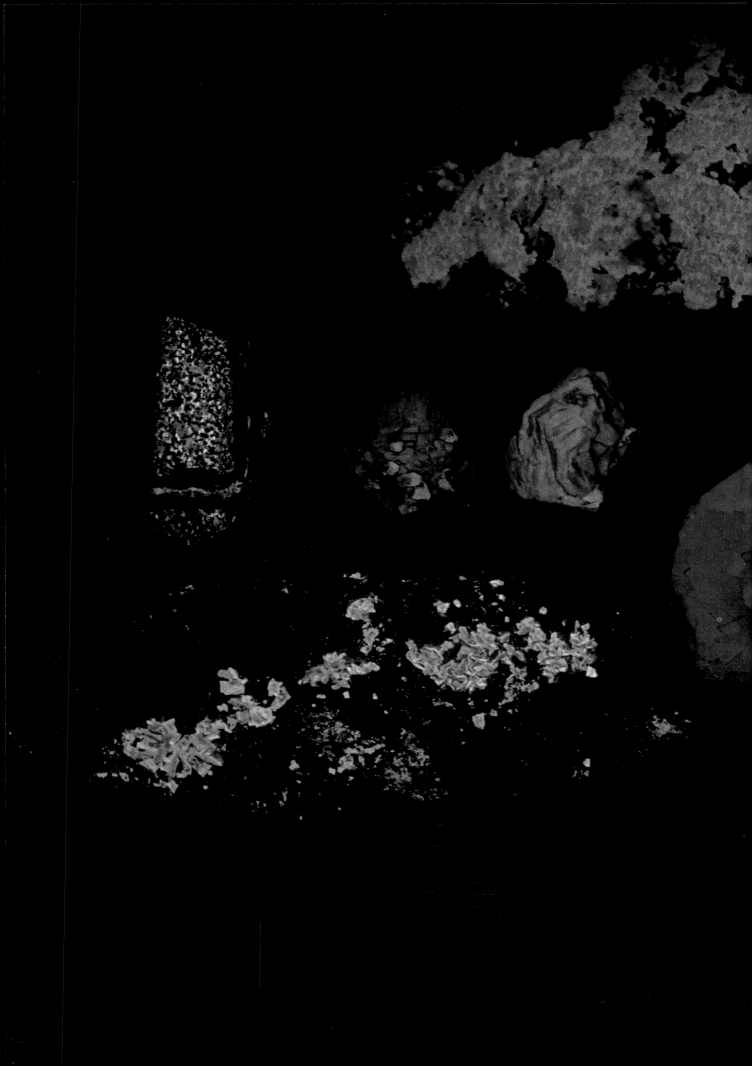

has a relatively fixed position and a fixed amount of energy within the atom. When the mineral is exposed to ultraviolet light some of the electrons absorb it, thereby becoming over-energetic and moving out of normal position. They immediately attempt to get rid of this extra energy and return to their original positions. The trapped surplus energy is given off as a tiny bit of light and all returns to normal. When this is done simultaneously by millions of atoms of the mineral, all the bits of emitted light added together produce the fluorescent glow.

Any human being who has ever tried to walk through a closed glass door or gazed through diving goggles at a spiny lobster in tropical waters must marvel at the phenomenon we call *transparency*. We are told that light travels through glass, water

and other substances, emerging intact on the other side. We are puzzled, however, as to how the light, bearing its images, can actually get through these objects. And if it can go through some solids and liquids, why not through others?

The fact is that light doesn't 'go through' anything. What happens is that light hits the atoms in these solids or liquids and, by its energy, starts them vibrating sympathetically. These vibrations are passed through the structure from atom to atom. If the atoms are properly aligned, the vibrations will move as in a row of falling dominoes. They are then emitted at the other side in the identical form in which they entered. The light does not trickle slowly through spaces between the atoms, as though finding its way through a maze. On the contrary, since the responsive vibrations are extremely rapid, the light travels at a rate close to the 670 million miles per hour at which unimpeded light travels.

Colour in minerals results from a complex series of interactions. Once we accept the fact that the passage of light through a mineral substance is subject to the way the mineral's atoms are arranged, it is easy to see that there might be alterations along the way. This tampering does not usually involve a change in wavelength; it is rather that the atoms may be selective about what wavelengths or colours they will permit to pass. If they are arranged so that only red passes, then red is all that emerges. Consequently, the mineral appears transparent, or partially transparent, red. If only green passes, the mineral will be transparent green. All the rest of the vibrations are absorbed in the labyrinthine arrangements of atoms. This, then, is a major source of colour in transparent minerals or, for that matter, in all transparent objects. If the mineral stops the passage of all light it looks opaque and black. Black is the absence of all visible wavelengths.

Sir Isaac Newton (1642-1727) founded mechanics and the science of optics, and discovered, among other physical facts, the colour spectrum. Legend—almost certainly untrue—holds that his theory of gravity was inspired by a falling apple.

The colour of a mineral, even if it is opaque, seen by light reflecting from its surface, results from a similar but opposite cause. Often, as a substance is struck by white light, some wavelengths are transmitted through it, some are absorbed, and some are not permitted to enter but are bounced back, or reflected. If the reflected wavelength is blue, the object looks blue. On the other hand the object can be red as we look at the light through it, since red could be the wavelength transmitted. Thus an object can appear to be different colours at the same time, depending on whether we see it by reflected light or by transmitted light.

Minerals are sometimes found to be different colours in different deposits. Fluorite, for example, is found at various mines in specimens of blue, green, pink, yellow, brown, purple and black. Absolutely pure fluorite is colourless. Any colour in it, apparently, must come from something added to it that is not typical of pure fluorite. Coloured minerals of this type are called *allochromatic*: the colour is due to traces of impurities or to defects in the arrangement of atoms in the structure. There are other minerals that have a characteristic colour not dependent on impurities. Sulphur, for example, is always bright yellow when pure. These are called *idiochromatic* minerals.

In addition to sifting and sorting the wavelengths of light, minerals are capable of *bending* it. This is fortunate, for otherwise we would have no lenses for telescopes, microscopes or eyeglasses, all of which depend on the bending of light to operate. Lenses are designed to bend rays of light, making them converge upon the eye in such a way as to produce the illusion of magnification. To prove this put a pencil on a slant halfway into a glass of water. The pencil looks bent where it enters the water. Light travelling from certain substances to others—such as air to water or water to glass or air to glass—bends. The amount of light-bending that takes place depends largely on what materials are involved. Bending is extreme

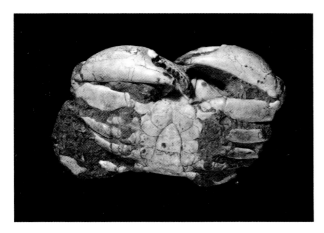

if there is a great difference in the density of the materials between which the light is passing. For example, there is more bending when light passes from air into glass than when it passes from water into glass.

Many crystalline mineral species bend or refract light in two directions at the same time. In such a case, a light beam is actually split into two parts, each being bent to a different degree. This is called double refraction.

Refraction and double refraction result from the way atoms in a mineral affect light pouring through it. As the degree of refraction differs for each mineral, it can be used as an aid in identification.

To say that all the reaction between light and minerals is due to the kinds of atoms present and their arrangement would be an oversimplification. There are certain other colour effects that are caused by structural defects, or by the presence of relatively large impurities. One of the more interesting of these processes is called *interference*.

Interference colours are usually explained by invoking images of thin films, such as an oil slick on a rain-wet street, which often produce a lovely liquid rainbow. A ray of light strikes a thin layer of oil at a particular angle. Some of the ray is reflected immediately. Some of it penetrates the film and,

in turn, is reflected from the contact surface between the oil and water. This second portion, travelling back through the oil film, continues on its way parallel to the first ray fraction. This means that the light waves in the two parts bouncing back are out of step with each other. The resulting combination is seen by the observer as a different mixture of wavelengths from the original ray or, by definition and appearance, a different blend of colour. The hue produced as these wavelengths interfere with each other depends upon the thickness of the oil film and the angle at which light strikes it.

The beautiful blue glow, or schiller (German for play of colours), exhibited by peristerite is caused by light interference of this kind. In this case the thin films are actually the layers of thin plates of which this feldspar mineral is formed.

Feldspar exhibits another light effect called *aventurescence*. In some samples of feldspar there are innumerable microscopic plates of the minerals hematite or goethite. These plates, often red-orange, are scattered throughout the almost colourless feldspar but are arranged parallel to each other. As light penetrates the feldspar, it is reflected at the same instant from myriads of these tiny, flat inclusions. Such feldspar is fittingly named sunstone or aventurine. Other mineral species, of course, can be aventurescent. One variety of quartz, for example, which contains tiny reflecting plates, usually mica, is also called aventurine.

Since the effects of interference and aventurescence are both found in certain kinds of feldspar, the two phenomena may occur in the same specimen. This phenomenon is called *labradorescence*, from its appearance in a calcium-rich feldspar called labradorite. It appears as large patches of iridescent blue and green colour with occasional reds and yellows. Most of this effect is due to the same light interfer-

Preserved for ages in sedimentary rock, fossils must be exposed by careful handwork. FACING, FAR LEFT: *Eurypterid from the American midwest.* LEFT: *Crinoids, same area.* ABOVE: *'Swimming crab' from Panama.*

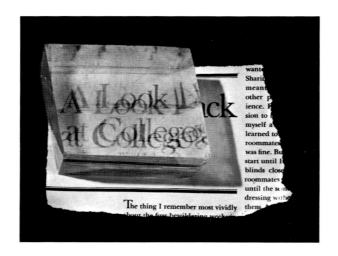

ence as is found in peristerite. Some of the colour flashes, however, are caused by the reflection of light from numberless platelets of black magnetite which are arranged as parallel inclusions.

RADIOACTIVE MINERALS: Legend has it that Henri Becquerel, professor of physics at the Museum of Natural History in Paris, accidentally discovered radioactivity in 1896 by inadvertently leaving some uranium ores on wrapped photographic plates in a drawer. This is a romantic story, but does Becquerel little justice. Actually, X-rays had been discovered and announced by Röntgen earlier that year. Becquerel undertook experiments of his own with a synthetic uranium compound—potassium uranyl sulphate—to see if he could produce X-rays by the action of the sun's rays on fluorescent materials. He thought he had done it because in one experiment his compound, placed in the sun on a covered photographic plate, fogged the plate as X-rays would have done. As sun was necessary to further experiment, during a spell of cloudy weather a batch of plates and the uranium compound were left together in a drawer. It was five days before the weather cleared. Though suspecting that the plates would be ruined after so long a delay, Becquerel, a careful scientist, developed them anyway, and found

to his surprise that the sun had not been necessary to cause the compound to fog the photographic plates quite badly. They had mysteriously become fogged while locked in the dark drawer!

Subsequently, he found that nothing he could do would stop, start or otherwise affect this strange radiation. The only requirement seemed to be that the element uranium or thorium had to be present in the substances he used to produce the penetrating rays. But unlike X-rays, Becquerel's strange rays were no good for taking pictures. Consequently they created little interest, even among scientists.

At about the same time, Marja Sklodowska Curie (1867-1934), intrigued by Becquerel's discoveries about radioactivity, began work on the radioactive substances in pitchblende as a project for her doctor's degree. Eventually, Madame Curie and her husband (who had been her physics professor at the Sorbonne) found that these strange penetrating rays had sources other than uranium and thorium. In 1898, they discovered polonium, which they named for the then non-existent Poland that was Madame Curie's homeland, and radium, named for the intensity of the radiations that this element gave off. Further experimentation demonstrated that these strange natural rays were far more penetrating than X-rays and far more deadly to living tissues.

Polonium and radium, interesting as they are, appear only as traces of impurities in other radioactive minerals. Usually when we speak of radioactive minerals we mean uranium minerals, because uranium is by far the commonest of radioactive elements. It is one of a group of elements that characteristically form highly coloured compounds; uranium minerals are bright red, orange, yellow and green. The only major exception is uraninite—uranium oxide (UO_2)—which is jet black.

The remarkable thing about radioactive minerals

ABOVE: *A calcite cleavage shows the mineral's excellent double refraction.* FACING: *Two optical phenomena as illustrated in a nineteenth-century book on natural philosophy.* TOP: *Refraction.* RIGHT: *Reflection.*

is not only what they do, but what happens to them after it is done. Uranium atoms that undergo the radioactive experience seem to disappear, to be mysteriously replaced by atoms of some other element, usually lead. For centuries, alchemists had worked unsuccessfully at the transmutation of elements, trying to convert common metals into gold. All this time nature had been busy transmuting uranium into lead, a truly remarkable feat.

For an atom of one element to become an atom of another, nature must tamper with the inner mechanism of the atom itself. A uranium atom has 92 particles called protons in its dense central core or nucleus, as well as 146 neutrons. At unpredictable intervals some of the neutrons are violently disrupted, upsetting the balance of things. Parts of the nucleus are then ejected at enormous velocities. These atomic projectiles have been labelled alpha (α), beta (β), and gamma (γ) radiation by physicists, and by the characteristics of the radiation have been carefully described. After the disturbance subsides, there are 82 protons left with 124 neutrons. This is an atom of lead. The results are always the same, but it is impossible to predict which atom will be transmuted next from uranium to lead.

For our purposes, one important fact is known. In five billion years, half of any given mass of uranium atoms will become lead. This time interval is called the half-life of uranium: in five billion years half of the uranium will be converted. The rate is as steady as the most accurate clock and cannot be changed.

This fact can be an invaluable guide for determining the age of rock deposits in the earth. For example: A specimen of rock is obtained that contains some uranium minerals. To determine exactly how old this specimen is, one must simply determine the amount of uranium and lead present, then calculate to see how long a time has been necessary to produce this quantity of lead in proportion to the remaining unconverted quantity of uranium. We then know that date when nature put the uranium in the rock or, in other words, the age of the rock itself. This method has dated some of the world's oldest known rocks, showing them consistent with the determination of astronomers that the earth was formed about five thousand million years ago.

All the remarkable mineral characteristics we have discussed so far have been demonstrations of energy being absorbed, changed or ejected. However, the wondrous appearance of many mineral objects such as concretions, geodes, stalactites, meteorites and pseudomorphs is actually the result of energies and conditions that created them in the first place.

CONCRETIONS: Roughly rounded masses of mineral, formed when some material is deposited around some kind of centre or nucleus, are called concretions. They are found embedded within an enclosing rock, not just resting in holes or cavities in the rock. Since they are usually harder than the rock surrounding them, they often remain intact even when their rock covering is worn away by erosion.

Some concretions began as rather stiff, gummy lumps that rolled around picking up leaves, shells, twigs and other debris. Eventually these lumps hardened and were buried or trapped in sedimentary rock beds. Other concretions were apparently formed by water seeping through sediments, dissolving some substances and later depositing this material in cavities in sedimentary rock. It is even possible for the concretion to grow in the sedimentary rock itself and not in a cavity, pushing the host material aside to make room.

Concretions occur in many shapes: spherical, ball-shaped, branch-like, flattened, oval, root-like. They can look like strings of beads, toy-balloon animals, human portraits, modern sculptures, solidified plants—almost anything imaginable. They vary in size from small, like the claystone concretions left by the Ice Age in New Jersey and southern New England clays, to specimens up to thirty feet long, found in the sandstones of Texas.

Concretions are usually composed of the minerals calcite, quartz, hematite, limonite, siderite, pyrite and marcasite, as well as clays. The beautiful green gem mineral variscite, an aluminium phosphate found in Utah, occurs as concretions formed in limestone. The brassy, metallic mineral pyrite is often found as concretions in coal beds.

GEODES: Geodes are sphere-shaped stones, usually at least partially hollow, and often lined with sparkling crystals or concentric layers of minerals.

Magnificent geodes make the area of Rio Grande do Sul in Brazil, and the adjoining part of Uruguay, notable sources of fine amethyst. In these regions, ancient volcanic rocks are filled with cavities which were produced by large steam bubbles at the time of formation. These bubble cavities later became lined with crystals of amethyst and agate deposited from solutions of seeping water.

From Rio Grande do Sul came the most fantastic geode known—thirty-three feet in length, sixteen and a half feet in width, and ten feet in height, with an estimated weight of seventy thousand pounds. Its interior was lined with tons of beautiful purple amethyst crystals, many of them several inches across, with glittering crystal faces. This geode provided many fine gemstones. Several sections were preserved intact, including a piece weighing four hundred pounds which is now at the Smithsonian Institution.

TOP: *Petrified wood from Clover Creek, Idaho.* RIGHT: *An insect trapped ages ago in flowing sap remains imprisoned in an amber fragment from the Baltic Sea area.*

Amethyst is a purple variety of the common mineral quartz, which is usually white or colourless. Quartz is the most common material found in the cavities of geodes. It appears in many colours, depending on what impurities have been gathered in during the lining operation and how rapidly this operation has taken place. Slow deposition, with relatively few impurities in the quartz, tends to leave the cavity lined with sparkling crystals of various colours and sizes. Rapid deposition will often fill the cavity with beautiful, varicoloured bands or patterns of extremely fine-grained quartz called agate.

Geodes are especially common in the limestones of the Mississippi and Ohio River Valleys. Large quantities of interesting and attractive geodes are also being found in Chihuahua, Mexico. Their contents include quartz crystals of amethyst colour, smoky colour, grey, white, lavender and brown, along with contrastingly coloured crystals of calcite, goethite and other minerals.

STALACTITES: Anyone who has seen a row of icicles forming on the edge of his roof has witnessed the birth of stalactites. With a change in temperature, these beautiful spears are gone forever. But fortunately nature also makes them elsewhere of durable minerals.

A common occurrence of stalactites is in the numerous limestone caverns scattered round the world. The Luray Caverns and the Carlsbad Caverns in the United States, the Cheddar Caves in England, the caverns at Postojna in Yugoslavia, the Jenolan Caves in Australia and many others are filled with rock icicles of all sizes and varied shapes. The variation in shape is caused by changes in the direction of water flow and in the rate of flow, producing curved and fluted, fat and short, long and thin, single or clustered groups of stalactites. On the

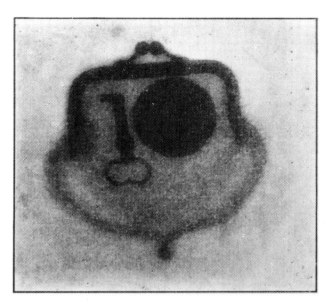

floors of these caverns stalagmites rise up to meet their pendant cousins, as the drops spill down to deposit more material below.

Stalactites need not be white. In some Arizona mines, impurities of copper minerals tint the stalactites a beautiful shade of green or bluish-green. Iron impurities can make stalactites pale yellow, brown or reddish-brown. Soldiers stationed at the not-so-solid Rock of Gibraltar used to occupy their long hours by carving various objects from the stalactitic onyx, banded and patterned in tan, brown and yellow, found in the limestone caves of that famous site. And although aragonite and calcite are the most common stalactite minerals, these unusual forms are sometimes composed of baryte, marcasite, limonite and malachite as well.

METEORITES: The passage of celestial objects through the earth's atmosphere is a phenomenon that has created widespread amazement since man first looked up at the heavens. The first observed meteorite fall from which actual samples are still preserved occurred at Ensisheim, near the French-German border. An eyewitness reported:

'On Wednesday, November 7, the night before St. Martin's Day in the year of our Lord 1492, a singular miracle happened: for between the hours of eleven and twelve a loud clap of thunder took place, with a long continued noise, which was heard at a great distance: and a stone fell from the heavens in the Ban of Ensisheim which weighed two hundred sixty pounds . . . it was a miracle of God: because, before that time, nothing of the kind has ever been heard of, seen or described.

'When this stone was found, it had entered the earth to a depth equal to the height of a man. What everybody asserted was, that it had been the will of God that it should be found.

'And when King Maximilian was here, the Monday after St. Catherine's Day of the same year, his Excellency . . . gave orders that it should be suspended in the church and that no person should be permitted to take any part of it. His Excellency however took two fragments: one of which he kept and the other he gave to Duke Sigismund of Austria.'

Contrary to orders, specimen samples were re-

FACING, TOP: *Henri Becquerel, discoverer of radioactivity.* LEFT: *A meteorite fall at L'Aigle, France, in 1803.* ABOVE: *Pierre and Marie Curie.* RIGHT: *A pinch of radium at one foot for one hour took the Curies' first radiograph.*

moved, and can now be found in several major meteorite collections as well as in the Ensisheim Church.

Early public discussion about meteorites and shooting stars had all the characteristics of present-day speculation about flying saucers. Thomas Jefferson, on hearing a report by Professors Silliman and Kingsley of Yale University concerning the fall of a meteorite in Connecticut in 1807, is reported to have said that it was easier to believe that two Yankee professors would lie than that stones should fall from heaven.

But stones do fall from the heavens, and in large quantities. Estimates indicate that four hundred million individual meteors penetrate the earth's atmosphere every day, and that twenty million of these become bright enough to photograph.

This means that space is filled with celestial debris swept up by the gravity of earth as it rushes on its way around the sun at 18.5 miles a second. The greatest part of this material is destroyed by the heat caused by the friction of its rocket-like passage through the air. But each year some few pieces survive, fall on the earth and are recovered. Four such recoveries in one year is considered remarkably good luck.

Meteorites are eagerly sought for scientific research, since they are our only samples of the universe that do not come from earth. Our current interest in space sciences has stimulated research on meteorites, and a good deal is now known of their composition and appearance. They are classified as iron meteorites, stony meteorites, stony-iron meteorites, and tektites.

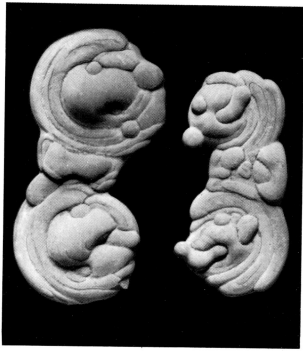

Iron *meteorites*, or siderites, are made of an alloy containing about 91 per cent iron and 8.5 per cent nickel, with traces of some other minerals often present.

A typical *stony meteorite*, or aerolite, has a black, glassy crust because its surface has melted during its scorching passage through the earth's atmosphere. When the crust is broken, a fine-grained stony interior is revealed. Stony meteorites are subdivided into two groups, chondrites and achondrites, depending upon their textures. The chondrites contain small chondri, which are round bodies composed of the minerals olivine or pyroxene. Achondrites do not have these bodies. Chondrites will have as much as 12 per cent nickel-iron, but achondrites only about 1 per cent. Other minerals present include enstatite, diopside, feldspar and olivine.

FACING: *Eroded remnants at Monument Park, Colorado, drawn from an early photograph.* TOP: *Four-foot concretions weathered out of shale near Tyrone, Pennsylvania.* ABOVE: *Clay concretions from Vermont— often called mudstones.*

Stony-iron meteorites, siderolites, contain about equal amounts of metal and stone. The stony part is usually olivine with some pyroxene, and is found as large rounded grains scattered throughout the whole sponge-like metal mass of this kind of meteorite.

Tektites are the subject of considerable theorizing. Not all scientists agree that they are, in fact, meteorites. In any event, tektites are small pieces of rounded glass varying in colour from light green to black, and of rather unusual composition for a glass. Tektites of various types are found scattered by the thousands over certain limited regions of the earth. Moldavites are a type found in Bohemia and Moldavia, Czechoslovakia. Australites are a variety found scattered over a particular area of two million square miles in southern Australia. The unusual composition of tektite glass, the lack of association with volcanoes and volcanic glass, the fact that the particles are scattered in particular patterns over certain areas, and finally their odd shape—all this suggests an extraterrestrial origin and rapid passage through the atmosphere.

The fall of meteorites has always excited man's

ABOVE: *Cascading pools in the archaeologically famous limestone cave of Domica in Czechoslovakia.* FACING: *Pumice, light enough to float, is formed from molten igneous rock of volcanic origin which has been made frothy by gas bubbles.*

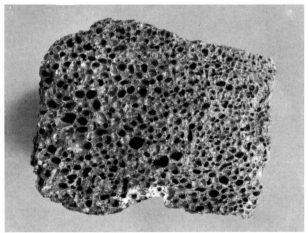

imagination—so much so that accurate observation is rarely made. Scientific literature is filled with reports of what people thought they saw rather than what they actually did see. Whenever real or imaginary meteorites fall, they are usually reported as being glowing hot. Later examination, however, sometimes shows that bits of grass pressed to the surface by the impact are not even charred. The fact is that meteorites pass through the atmosphere so rapidly that only their surface skin heats up, glows, melts and sloughs off. The interior does not even have time to warm up from its frigid passage through the incredible cold of space. Freshly fallen meteorites have even been found with a coat of frost formed by condensation of moisture on the cold surfaces.

Perhaps the best-known hole in the world was made by the bullet-like impact of a meteorite: the Canyon Diablo meteor crater near Winslow, Arizona. This old crater, formed perhaps thirty or forty thousand years ago, measures almost one mile across the rim and is almost six hundred feet deep. Even after thousands of years of erosion, the crater remains, and thousands of pounds of scattered meteorite fragments have been recovered from as far as eight miles away from it.

In early 1947, the hot, compressed shock wave of air from a falling meteorite two hundred miles north of Vladivostok, Russia, laid waste many square miles of forest. The blast of air shattered solid rock, tore up large trees and tossed them several miles away, and gouged out dozens of craters, some almost one hundred feet wide and fifty feet deep.

And yet the fall of meteorites may sometimes be gentle. Dr. George P. Merrill, formerly of the Smithsonian, reports in *Minerals from Earth and Sky* an unusual fall near Hessle, Sweden, in 1869,

Varied specimens result from the sedimentary processes of mineral-bearing solutions. BELOW: *Fossil brachiopod replaced by pyrite, from Ohio.* RIGHT: *Sand inclusions in azurite, from France.* BOTTOM: *Coral replaced by chalcedony, Florida.* FACING: *From Argentina, cross-section slices of rhodochrosite stalactites.*

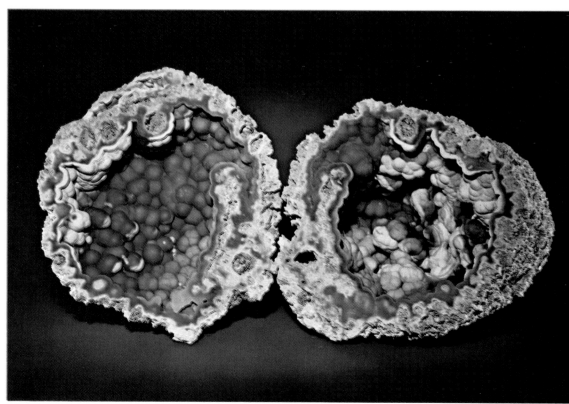

122

when 'the stones fell in great numbers. . . . Although so brittle as to be crumbled between the thumb and finger, few were broken by force of impact, and one which fell on the ice rebounded without rupture.'

It is not widely realized that there is also, from the heavens, a constant gentle sifting down of the fine ash remains of all the millions of meteorites destroyed by our atmosphere each day. In fact the earth proceeds along its orbit like an enormous vacuum cleaner, sweeping up multitudes of celestial objects.

PSEUDOMORPHS: Australians can find petrified wood within easy reach of Hobart and Brisbane and on Garie Beach near Sydney. In America the Petrified Forest of Arizona is a famous attraction,

and petrified wood has become so popular that the government has had to set up regulations controlling its removal from federal property.

Petrified wood is extraordinary material. It looks very much like the wood that it once was, which has now been completely replaced by stone. Some of the handsomest petrified wood specimens are found in the Petrified Forest of northern Arizona. Millions of years ago, floods swept these trees down from the hills year after year. In the process the trees were stripped of leaves, bark, roots and branches, and then buried in muds consisting mostly of volcanic ash. Volcanic ash is rich in quartz, and also contains some iron and manganese. Water, seeping through the mud, dissolved the

minerals from the volcanic ash and carried them to the buried tree trunks, where they slowly and delicately permeated the wood spaces. Eventually the minerals entirely replaced the cell structure of the wood, though it was not destroyed in form. Iron and manganese impurities added vivid dashes of bright red, orange, yellow, brown and black. Millions of years later the erosion of the burial grounds gradually brought to light these logs, now forever preserved as brightly coloured stone.

It is not only living things like wood that can be replaced by mineral species. Minerals may actually replace each other without a change in external form. Such replacements are called pseudomorphs.

From the moment minerals are produced by one

124

TOP: *The ancient, mile-wide Canyon Diablo meteorite crater, Arizona.* ABOVE: *Meteorite sample sawed for study.* FACING: *Two meteorites, from Thiel Mountains, Antarctica, and Toluca, Mexico, contain olivine and a nickel-iron metal alloy—combinations not found on earth.*

process or another, they are subject to chemical attack and alteration if there is change in their surroundings. Some minerals resist change and are relatively stable. Others change so easily on exposure to heat, light and moisture that it is impossible to preserve them. Changes can be of several types. The simplest is a loss of part of the mineral's constituents, usually water. Water escapes from some minerals into the air merely on standing. In other minerals, part of the constituents may be removed by the dissolving action of solutions that nature sends flowing by. Constituents may be as easily added as removed by the same solutions.

Sometimes whole groups of a mineral's crystals may be dissolved away, leaving in a rock cavities or molds that preserve their exact shape. Later these holes or molds become filled with quartz, calcite or some other mineral species brought in by solution. This process makes perfect casts of specimens, which have the shape of the original mineral but none of the other original characteristics. This explains the origin of their Greek name pseudomorph, *pseudo* meaning false and *morph* meaning shape.

There are many kinds of pseudomorphs known, including precious opal after (which means replacing) glauberite, quartz after fluorite, pyrite after shells and other fossils, clay after feldspar, and malachite after azurite.

LODESTONE: Marco Polo, in accounts of his voyages to ancient China, describes the use of special stones as direction finders. The ancient Chinese knew that certain pieces of iron ore had the ability to align themselves in a certain direction when held in suspension. And by the fourteenth century, when Geoffrey Chaucer mentioned lodestones in his 'Shipman's Tale', magnets had become an important piece of navigational equipment.

Lodestone—originally 'lead stone' because of its use as a direction finder—is a peculiar variety of the mineral species magnetite, an oxide of iron (Fe_3O_4). It attracts bits of iron and steel with a strange force, which even now is not completely understood. This same force surrounds the entire earth, making our planet itself a very large magnet. It also surrounds any electric wires through which a current is flowing, making it possible to create powerful electromagnets with thousands of coils of wire. These electromagnets are used in turn to endow pieces of iron, steel and other metal alloys with the same magnetic characteristic by exposing them to the strong magnetic force for a period of time.

On this and the facing page, a group
of minerals that characteristically
display light and radiation effects.
ABOVE: *Radioactive, fluorescent
autunite from Spokane, Washington.*
RIGHT: *Labradorite from Labrador,
turned (FAR RIGHT) to display
schiller.* FACING, LEFT: *Iridescent
limonite from Mexico.* TOP, RIGHT:
*Radioactive curite from the
Congo.* BOTTOM, RIGHT: *Radioactive
uranophane from New Mexico.*

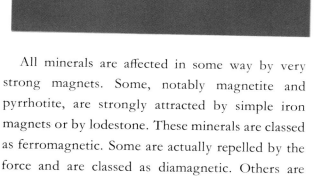

All minerals are affected in some way by very strong magnets. Some, notably magnetite and pyrrhotite, are strongly attracted by simple iron magnets or by lodestone. These minerals are classed as ferromagnetic. Some are actually repelled by the force and are classed as diamagnetic. Others are attracted to varying degrees, and are classed as paramagnetic.

Because each species is attracted or repelled to a specific degree, it can be magnetically separated in the laboratory, in industrial and mining procedures, and in prospecting. A magnetometer—an instrument that registers variations in the earth's magnetic force—can be carried or flown over large areas to detect the presence of iron-ore deposits by recording variations in the earth's magnetism.

INCLUSIONS: When the Seven Dwarfs placed Snow White in a crystal casket to preserve her from harm, they were duplicating a long-established natural process. Almost every time nature grows a crystal, she encases in it a variety of objects, called inclusions. They are scientifically important because they are evidence of the temperatures, pressures, composition and other characteristics of the environment in which the mineral formed.

Not only solids but liquids and gases are often trapped during crystal growth. The difference between the white, opaque variety of quartz called milky quartz and the clear, glassy variety called rock crystal is caused by multitudes of tiny bubbles of liquid trapped in milky quartz.

Frequently the inclusions will be of more than

one kind, each called a phase. One of the most amazing sights to see under a microscope is the two-phase inclusion of a tiny bit of carbon in one of the liquid-filled cavities found in quartz from Herkimer County, New York. The carbon can be seen jittering around in an erratic dance, bombarded this way and that by the ever-moving liquid molecules. This is called Brownian Motion, after Robert Brown, the botanist who first reported it in 1827.

THE LIQUID MINERALS: People usually think of petroleum when liquid minerals are mentioned. The fact is that petroleum is not a mineral by our definition, since it evolves from living things through an organic process. However, there are two liquids—one common, one rare—that are actually minerals. The common one is water, a non-metal, and the rare one is mercury, a metal.

Water, of course, is very familiar in its solid state, either as masses of interlocking crystals (ice) or as accumulations of individual crystals (snow). Perhaps the most interesting and fortu-

nate characteristic of solid water is that it is less dense than liquid water and therefore floats on it. Consider what would happen if water at the surface of a lake froze and sank because it was more dense. Sitting on the bottom, it would soon be covered by more sinking ice. Eventually the whole lake would be filled with solid ice, perhaps never to thaw again. As it is the ice layer floats, insulating the water under it from exposure to the low temperatures of the air above.

Liquid mercury is popularly called quicksilver because of its silver colour, and because a globule of the liquid rolls rapidly over any surface without wetting it. Quicksilver is thirteen and a half times as heavy as water. It readily forms solutions, called amalgams, with gold and silver. The dentist takes advantage of this in filling teeth, since silver could never be worked into the cavity in its hard metallic form.

The processing of gold ores also relies heavily on mercury's ability as a solvent. The crushed ore is passed over mercury, the gold in the ore becoming an amalgam with the mercury. The amalgam is then removed and the mercury boiled off, leaving behind a spongy mass of quite pure gold. Mercury's amalgam-forming property was well known to the ancients; Pliny described its use in gilding copper and silver.

Very little liquid mercury is found in nature. Almost all commercial mercury is mined as cinnabar, a compound of mercury and sulphur (HgS). Because it is bright red, cinnabar has been used for centuries as a pigment and is still one of the best. The mercury itself is recovered by roasting the cinnabar and driving off the sulphur.

CATACLYSMIC ROCKS: There are times when portions of the earth's crust are subject to violent disruptions or cataclysms, which may produce unusual objects not formed under ordinary circumstances. Three or four rock relatives of quartz illustrate this kind of occurrence. Perhaps the strangest of these is the fulgurite.

Fulgurites are twig- and branch-like, amorphous structures, some of which are several feet long. They are formed when a bolt of lightning strikes a bed of quartz sand. The quartz is fused by the passage of the high-intensity electrical discharge. If the fulgurite is carefully dug out, it is found to be a perfect glass picture of a highly branched electric spark.

The cataclysmic eruption of volcanoes provides other examples of such mineral wonders. Lava is sometimes full of dissolved gases under pressure. When the pressures are relieved, the lava froths and then solidifies. The result is a glass, related to quartz, that is full of tiny gas bubbles. It is called pumice, and is such a light substance that it will float on water. If the gas bubbles in the solidified froth are quite large, the rock assumes a honeycomb appearance and is called scoria. Scoria, too, is light enough to float.

Even more unusual is Pele's Hair, named after Pele, the Hawaiian goddess of the volcano. Specimens of this lava actually look like silken strands of hair. It is formed when drops of liquid lava are spewed out, stretched, or spun into long glassy hairs, which the wind then tangles together in intricate masses.

The wonders of the mineral kingdom appear limitless. With science and industry pressing for further exploration, and with increasingly sophisticated tools and techniques, it seems likely that earth scientists will frequently be adding many further discoveries to the remarkable phenomena that have already been revealed.

FACING: *Magnetism, shown here in magnetite or lodestone, is a phenomenon not yet fully explored.* ABOVE: *Fulgurite, formed when lightning strikes quartz sand.* BELOW: *Tektite from South Australia.*

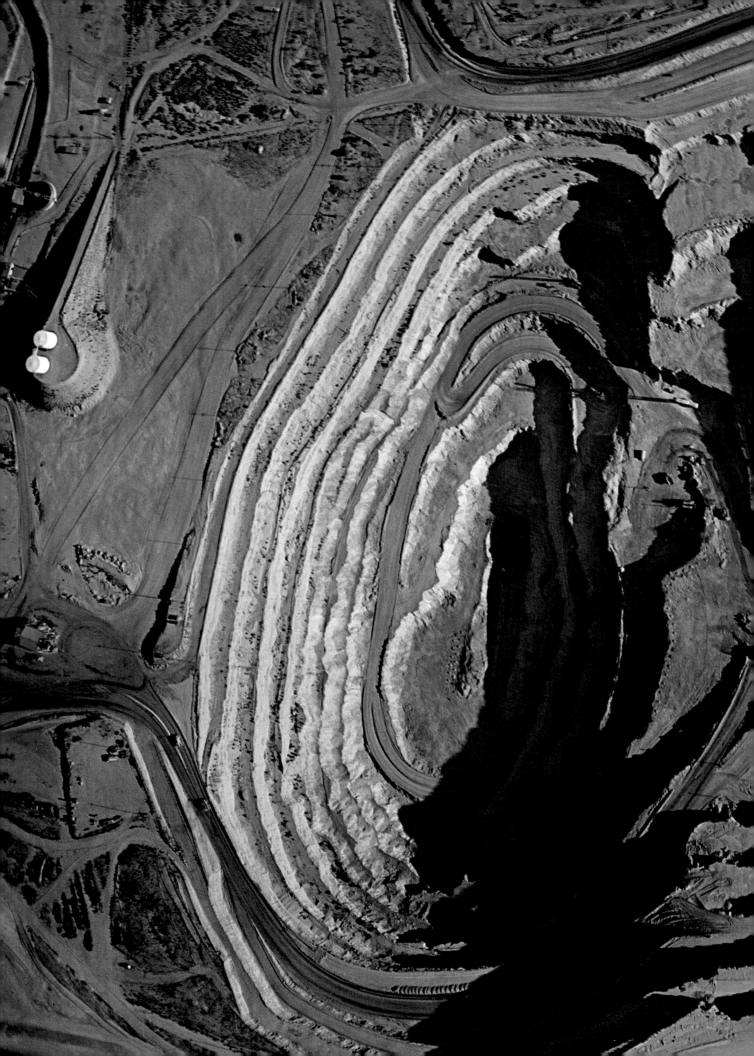

In Search of Treasure: Classic Sites

5

Every country of the world has its mining districts, and each of these has its own methods and history. Six of the districts, swinging around the globe from California to Australia, illustrate the variety of man's difficulties as he presses the search for mineral treasure. Each of these areas is a classic site for the recovery of some mineral substance as highly prized in today's brisk commercial trade as it was in centuries past.

SOUTH AFRICA'S DIAMONDS: The Boers, who trekked northward in 1850 to escape the dominant English influence in the original Cape Colony at the tip of South Africa, crossed the Vaal River into the Transvaal to carve out new farms in peace and tranquillity. They found themselves, instead, over-run by the would-be miners of the diamond rush of 1870.

It was triggered when, in 1866, young Erasmus Jacobs found on his father's farm a pretty pebble, which he gave to his sister for a game they called 'five stones'. This pebble turned out to be a 21.7-carat diamond. Bought by the colony's governor, it was displayed in Paris as the Eureka diamond—and the rush was on. Hordes of fortune seekers, fresh from the gold fields of California, Australia, New Zealand, swept into Transvaal seaports and spread out over the whole area.

At first the Dutch farmers and native Africans fought them off. But as new discoveries were made—among them the 83.5-carat Star of South Africa, another 'pebble' picked up downstream from the Jacobs farm—the tide of invaders became over-whelming. The diamond-mining industry had come to stay.

Some of the greatest finds were made by chance. In the crown colony of Griqualand West, a servant boy set to digging as a punishment discovered diamonds on the campsite of the Red Caps, a small prospecting group. Almost overnight, the site became Kimberley—named after Britain's foreign secretary.

Another farm in the territory, bought from the De Beers brothers, became the site of a major mine that was eventually to produce a fortune in diamonds. Diamonds were also found during this period on a farm to the south near Jagersfontein. Soon after, in the bottom of a cattle pond—Du Toits pan—some farmers found a heavy salting of diamonds. All of these occurrences gave their names to the great mines of today—Dutoitspan, Bultfontein, De Beers, Kimberley, Jagersfontein and Premier.

The diamond mines are located in two kinds of diggings—river gravels and 'dry' diggings. The river gravels were the hunting grounds of the first prospectors—from Hopetown, along the Orange River to its junction with the Vaal River, and also up the Vaal as far as the town of Potchefstroom.

In these locations, diamonds occur because of millions of years of erosion which have torn away thick layers of the surface of Africa. Thousands of feet of surface have been carried off toward the sea, producing extensive deposits of diamond-bearing gravels and ocean-bottom deposits along the coast of Africa.

Later, prospectors found diamonds in dry diggings on various farms, most of them on a plateau between the Vaal and Modder Rivers near Kimberley. Here the diamonds are found in deposits of rock of igneous origin called kimberlite, which goes down to considerable depths and is the filling in the necks or pipes of a number of extinct volcanoes.

The deepest mine in any of these pipes is the Kimberley, which goes down 3520 feet. The kimberlite and diamonds here go still deeper, but mining at this great depth becomes uneconomical. The largest of the pipe mines is the Premier, located near Pretoria in the Transvaal. On the surface it

PRECEDING PAGES: *Roads outlining the great terraced pit at the Mission Mine, a young open-pit mine in an enormous Arizona body of low-grade copper ore.*

covers eighty acres, and it extends to a depth of several hundred feet. The operation at the Premier is quite modern and it will someday go down to considerably greater depths.

Most of the pipes are barren of diamonds. Even in the rich pipes, four to six small diamonds in a cubic yard of ore is a good average. Only large-scale mining can work such ores profitably. The original rock of which most of the kimberlite consists is called peridotite, and is composed of the mineral species olivine with some bronzite, chrome diopside and other substances. At the top of the pipe, exposed to the atmosphere, the kimberlite decomposes and becomes yellowish and crumbles easily.

It was in this 'yellow earth' that the first pipe-seated diamonds were found. Hauled out and dumped on 'floors'—flat cleared patches of land— the kimberlite weathered and decomposed still more, and was then crushed and washed to remove clay and other lighter materials. The heavy concentrates that remained were run over a grease table. Diamonds stuck to the grease, while the other minerals were shaken and washed along their way. Later diamonds were removed, cleaned and sorted for sale. Almost the entire process was accomplished with hand labour and muscle power.

Eventually, miners reached the bottom of the yellow earth in the pipes and assumed it was the end of the operation. Discovery that there were diamonds in the underlying 'blue earth' renewed mining efforts. Blue earth is also peridotite, but

An early nineteenth-century drawing shows basalt columns in Westphalia, Germany, originally upright, tilted by the earth's crustal movement. Lush vegetation springs from the covering of eroded volcanic soil, which is usually rich.

unexposed and unweathered to its yellow form. Exposure on the weathering floors remedied this situation.

The history of the most powerful force in today's diamond market—the mammoth De Beers diamond cartel—begins in the diamond rush of 1870. It was the custom of the diggers to make local agreements about their claims. Usually, no man was allowed to operate more than one claim at a time, and a claim was only thirty-one feet square. With miners working on top of each other, the mining areas became a great sea of churned-up earth in various stages of excavation and collapse. Slides and cave-ins, difficulty in moving off the loads of ore, water-seepage problems, and squabbles over conflicting claims created a state of chaos.

Into this scene came young Cecil Rhodes, eighteen years old. At first he only rented pumps to hold back the troublesome water seepage. But soon he began to buy up claims at the De Beers mine. By the time he was twenty-eight, he was the president of De Beers, and a rich man.

At the Kimberley, an enterprising promoter named Barnett Barnato had gained control and had handled the hazard of great shale slides by sinking tremendous shafts and converting to underground mining. But both Kimberley and De Beers suffered because the financial foundation of the industry was unsteady. Every time floods or slides threatened, the claim holders would panic and sell diamonds to cover anticipated losses, thus ruining the market.

Rhodes and Barnato both felt that a monopoly control of the big mines was necessary to stabilize the market. In the tremendous financial battle that ensued, Rhodes finally won because the Rothschilds were behind him. By 1889 De Beers Consolidated Mines Limited had its monopoly in diamond pipe mines.

During this period, Sir Ernest Oppenheimer appeared on the scene as a diamond buyer, moved into gold mining, and organized, with English and American money, the powerful Anglo-American Corporation. This company wisely began to buy up diamond mines north of the Orange River and alluvial deposits in Alexander Bay. Watching this activity, the De Beers Company in 1926 asked Sir Ernest to become a director. By 1929 he was chairman of the board and had sold the entire Anglo-American diamond operation to De Beers.

Losses through theft at the mines had been another sizable problem; at this time, the Illicit Diamond Buying Act was passed, making it illegal to buy diamonds except from licensed dealers. Also, security-tight compounds were established around the mines. All native labourers were carefully searched before leaving the compounds. With the leakage blocked, the monopoly and control of diamond sources was now complete.

AUSTRALIA'S OPALS: The opal region of south and east Australia is a tremendous spread of uninviting, primitive terrain. Thus far only a handful of opal-producing districts has been found, but there is every promise that more discoveries will be made. Isolation from civilization, lack of water, and almost unbearable temperatures deter most would-be miners. The erratic, unpredictable occurrence of precious opal has prevented development of large-scale mining companies using modern techniques. Thus, the comforts of civilization that usually flow into wealth-producing mining areas have not yet come to the opal diggings of New South Wales, South Australia and Queensland.

Common opal, with no dazzling display of fire, is found almost everywhere. But precious opal is

FACING: *In an ancient Japanese gold mine, the shaft follows the ore vein. Modern shafts go straight down.*

rarer than diamond. Hungarian opal mines yielded poor-quality gem material for many years, but not until the discovery of Australian opal was the active opal market created that now exists.

All through the accounts of discoveries in Australian opal country there runs a refrain of the unspeakable hardships suffered by early prospectors. Many of the mines are located in the 'outback' or the 'never-never' as it is called locally. This is the region through which, in one place, the Trans-Australian Railway runs for over three hundred miles in an absolutely straight and flat line and daytime temperatures average over 100 degrees. Only about five inches of rain fall each year. Drinking water must sometimes be transported many miles. Added to all this, the early prospectors and sheep-herders were often attacked by the aborigines.

In 1875, the Aladdin Opal Mine was founded in the Thackaringa Hills by an itinerant miner named Paddy Green. Some of his gorgeous flashing opals—locally called 'Yowah nuts' because they were first found near the Yowah homestead—were taken to the gem dealers of Hatton Garden in London by a Queensland dealer named Herb Bond. Though London was then the centre of the world gem trade, the only opals the dealers had ever seen before were the inferior ones from the old Hungarian mines. Bond's specimens were so good that the dealers were certain they were artificial. He found no buyers.

Meanwhile, new discoveries were made, such as the boulder opal at Quilpie in Queensland. Here the opal occurred as coatings and as filling in fractures in brown boulders of concretionary iron oxide and quartz. Quilpie also produced a number of fine large pieces, the largest of which was a tree limb, found in 1898, which had been replaced by precious opal. Reportedly, this limb measured ten feet long by one foot in diameter. It was unfortunately broken up for cutting.

With only sporadic mining proceeding in the Yowah and other opal fields, the opal market remained limited to Australia. It was in this period that Tully Wollaston, a young surveyor, formed a famous partnership with an Adelaide lawyer, Dave Tweedie. Having obtained financing, Wollaston set out by horseback and camelback across the desolate wastes to find a miner named Joe Bridle, somewhere north of Quilpie, who was rumoured to have some fabulous opal. The difficult seven-hundred-mile gamble was successful; Wollaston returned with the gems.

When he took samples to London, in 1889, Wollaston met the same difficulties as Bond had years before. But he finally persuaded Hasluck Brothers to merchandise some cut stones in their shops in London and in Maiden Lane, New York. The venture paid off; the opal's beauty quickly found a market. Demands and orders began to come in. Wollaston and Tweedie now had a firm in operating condition, and the Australian opal industry was launched.

The famous White Cliffs mine got under way when, just short of seven hundred miles northwest of Sydney as the crow flies, a group of kangaroo hunters stumbled on opal in 1889. They brought pieces of the colour-flashing stone to the Mining Register in the nearest town, Wilcannia, sixty miles to the southeast. At this point a miner named Ted Murphy entered the picture, helping them to develop their mining operation. Murphy, from his experience at handling opal at White Cliffs, knew quality and prices and supply problems as well as anyone in the field. He later joined Wollaston and Tweedie, and eventually became the largest and most talented buyer of opal in Australia.

TOP: *Enormous low-grade deposits in east-central Arizona rock are important copper reserves.*
RIGHT: *Native copper from Michigan's Keeweenaw Peninsula, mined by Indians long before Europeans arrived.*

The first White Cliffs shaft went only to twenty feet because it was believed that this was the bottom of the deposit. Since that time many other and deeper shafts have been opened, much fine opal has been found and sold, and this opal field is now known to be at least five miles long and two miles wide. Many of the mines have exposed opalized shells and other fossilized animal remains. These finds confirm that this part of Australia, millions of years ago, was a vast ocean basin. Presumably, opal deposits can be found anywhere in the basin. By 'miners right', each miner was permitted to cover only one hundred square feet of ground in his operation. It is conceivable that with the persistence of primitive mining techniques, plus the forbidding nature of the country, the entire ancient ocean basin may never be completely searched, mined, or exhausted of its opal.

In the desolate country northeast of White Cliffs, a sheepherder grazing his flock was caught in a sweeping thunderstorm on a small rise. The hill rose only thirty-five feet, but this was sufficient for the entire close-packed flock to be wiped out by a single bolt of lightning. From then on the place had a name—Lightning Ridge.

At this spot, near the New South Wales side of the border with Queensland, opal was discovered in 1905. This opal differed from all the other kinds

then known because its background colour, in which the typical opal fire danced, was so dark. It immediately became known as black opal, and—though Sydney gem dealers rejected it as too dark, and worthless—it was so extraordinary that a flood of miners poured into the area. The short-sighted gem dealers found history and the gem market passing them by; fifty thousand acres of the area have since been declared an official mining field. Actually, less than 5 per cent of it is currently being worked.

Mines at Lightning Ridge are typical of those elsewhere in the opal fields. All the work is done with pick and shovel. The shafts go down about

twenty-five feet, through gravel, sandstone and quartzite layers, before the opal-bearing white clay layer is reached. They are about two and a half feet wide by about six feet long, just big enough for a miner to descend and swing a pick. Usually the mine is a two-man operation—one in the shaft and one to hoist the opal-bearing clay up the shaft by windlass and bucket. Sometimes a third and fourth are employed to sift the clay through a screen so that no scrap of opal is lost. 'Nobbies'—small round and almond-shaped masses of opal—are the final product. Although the area was discovered in 1915, it is still productive, and as recently as 1946 an aborigine discovered similar opal over eighty miles away.

The names Yowah, Quilpie, White Cliffs, Lightning Ridge, and the equally famous Coober Pedy, Eulo and Tintabar add romance to the discussion of opal, as Kimberley and De Beers do to diamond. In 1930 two boundary riders, out checking grazing fences, discovered a deposit that added the name Andamooka to this illustrious list. Only one hundred forty miles north of Port Augusta in South Australia, and not too far from the well-known Australian rocket firing range at Woomera, this has become a productive field.

There is no doubt that this is not the last of the big discoveries. The world market for good opal is large and active and prices have gone up dramatically, a fact that is certain to stimulate the search. Obviously, great portions of the potential opal-producing country in Australia have not been tapped at all and new discoveries are bound to follow.

FRANKLIN ZINC: The Dutch miners who worked the Delaware River Valley for copper ores about 1640 thought they had struck it rich when they came to the place now known as Mine Hill in Sussex County, New Jersey. The exposed brownish-red ore looked like the copper ores they knew so well.

ABOVE: *Open-pit mining was typical of early diamond mines like Kimberley. Here diamonds were found— and are still found—in deposits of rock of igneous origin called kimberlite.*

Mine workings were established by about 1650, but that was the last heard of the Dutch copper ventures at that location. For reasons that the Dutch didn't understand and that were not to be cleared up for another one hundred and fifty years, the red ore yielded no copper.

The area later passed into the hands of William Alexander—who preferred to be known as Lord Stirling though his claim to the title had been denied, apparently because of his active role as an American Revolutionary officer. Like the Dutch, he tried to make use of the red ore, and sent a shipment of it to England to be smelted. The smelting never took place, but the shipment probably accounts for certain specimens found in old mineral collections in England. Undaunted, Lord Stirling erected his own furnace at Franklin, hoping to smelt his plentiful black material, thought to be iron ore. The furnace never smelted any ore and in time fell into ruins.

By 1816, when the property came into the hands of

Dr. Samuel Fowler, there was strong evidence that the Mine Hill ores just weren't copper and iron. But what were they? Samples of them, making their way about the scientific community, were examined with curiosity, interest and no conclusions.

The mystery was solved in the first volume of the *American Mineralogical Journal*, published in 1814. In an article entitled 'Description and Chemical Examination of an Ore of Zinc from New Jersey', the red-brown mineral, zincite, was shown to be an ore of zinc and not of copper.

This paper was only the first of many. For years Dr. Fowler and his son encouraged scientists to continue studying the materials that were being recovered from the mines. A steady stream of mineralogists was entertained by the Fowlers and given specimens for study. The names of many important mineralogists from all over the world have appeared on a long line of scientific papers, which have been published steadily since Fowler's time.

Dr. Fowler tried repeatedly to exploit the deposits commercially, but it seemed hopeless. In 1850, control of the property passed on to the New Jersey Zinc Company. A number of mines sprang up, which in 1897 were consolidated and modernized. The sloping Palmer shaft, fifteen hundred feet long, was sunk to the base of the ore body, and from 1909 on, all ore from everywhere in the ore body was brought up through this shaft.

Actually, there are two ore bodies. One is at Mine Hill near the town of Franklin, the other three miles away near the town of Ogdensburg. Both are very extensive and are shaped in enormous troughs, rather like the two ends of a canoe. The geological origin of the ores is still somewhat controversial. It must have been a unique combination of geologic events, because there are no other deposits like them anywhere in the world. They sit in a region of coarsely crystalline white marble or limestone called Franklin limestone. In the limestone itself are some unusually fine mineral specimens for the collector. None of these contains zinc or manganese, which is strange since the ore bodies are rich in both. The Franklin limestone itself has been mined commercially.

At first the ore seems very simple in composition. Seventy per cent is made up of the minerals franklinite (an oxide of zinc, iron and manganese), willemite (a zinc silicate), and zincite (a zinc oxide). Twenty-five per cent is calcite, calcium carbonate. The remaining five per cent is an extremely numerous and interesting suite of well over one hundred and fifty different species. Many of these were first discovered at Franklin; many are not known anywhere else in the world, and much of the material has yet to be identified. Among the recognized minerals are very rare lead silicates—margarosanite, barysilite, nasonite, roeblingite and larsenite.

Investigation of these minerals has now extended over almost a century and a half. Many collections have been made of Franklin specimens all during the operation of the mines. Most of the important ones are in major mineral museums, still available for study.

These remarkable ores have been put to many uses. Franklinite, the most abundant, can be separated easily from the rest of the ores because it is magnetic and can be pulled out of the crushed mixture. The franklinite is then converted to white zinc oxide and to spiegeleisen. Zinc oxide is used as a non-staining whitener in paints and in the manufacture of certain rubber products. Spiegeleisen is an iron-manganese alloy used, especially in the Bessemer process, for making manganese steels. The willemite and zincite from the ores are kept together during ore separation and end up as spelter, or crude zinc, for use in various industries.

As the Franklin ores were worked, it was discovered that some of them were fluorescent, glowing in bright colours under ultraviolet light. This permitted easy separation of the waste material from the ores by hand, on picking tables, as it came out of the mine. The need for better methods of producing ultraviolet light specifically for this purpose so stimulated research that it led eventually to the present range of excellent light sources now available to industry and collectors.

The half upcurving mass of ore at Mine Hill, known earlier as Franklin Furnace and now commonly as Franklin, has been mined out and operations are closed down. However, over in Sparta Township the mine is in full operation. The story of the resistant Dutch 'copper ores' is far from fully told.

CALIFORNIA'S GOLD: The United States still

Some occurrences produce extraordinary specimens.
FACING, LEFT: *Anhydrite from Aussee, Austria.*
RIGHT: *Adamite from Mapimi, Durango, Mexico.*

Australia, not fully exploited, remains a promising hunting ground. ABOVE: *The 1851 gold rush drew hopefuls from around the world.* RIGHT: *At famous Lightning Ridge, an expert cuts and polishes an opal.* FACING: *Sapphire-bearing gravel is screened at Rubyvale in Queensland.*

has vast hoards of gold—in the ground where nature put it. Mine operators insist that far more gold remains in California, awaiting mining, than has already been dug out of the ground there.

There had been sporadic discoveries of gold in many parts of California before 1848, when it was still part of Mexico. Indians would occasionally bring samples to the Spanish missions. Richard Henry Dana, in his biographical *Two Years Before the Mast*, wrote that every ship calling at West Coast ports would bring gold dust back to the East. Of course, the territory belonged to Mexico, but when the Mexican War ended in 1848 with the Treaty of Guadalupe Hidalgo, which gave California to the United States, the lid was off.

The pot had almost boiled over prematurely. Just ten days earlier, gold had been found where a man named John Sutter was building a mill. As workman after workman found additional samples, word travelled, and nearby towns were depopulated by a stampede to the area. The news spread to the Oregon Territory, Mexico, South America, and abroad—and the tremendous rush of gold seekers toward California was under way.

Reaching California, the new settlers found an undeveloped wilderness with no government, no laws, no taxes, no comforts—but with unlimited

opportunity. They fanned out along a two-hundred-mile stretch in the Sierra Nevada mountains and set up their mining claims and tent communities on the slopes. Many struck it rich quickly—some by accidental discovery, others by very hard work.

Most of the gold was recovered by panning: separating the gold from the sands and gravels by washing with water. The heavy gold settled out and the lighter waste material was carried off. The panning was often done right in the stream beds. Where gravels were mined and washed at some distance away from the streams, long ditches were dug to bring the water to them.

A standard item of equipment was the 'rocker', a three-foot-long box set on rockers, with wooden cleats nailed at intervals on the inside bottom. Gold-bearing sand was shovelled in at one end, while a current of water ran through it. The heavy gold sank and settled on the cleats; the lighter gravel was washed along and out the other end. Sometimes, to give the gold a longer distance in

which to settle, the rocker was abandoned for a 'long tom'. This was a long trough, or several joined together, with no rockers but with a similar series of cleats in the bottom.

Hydraulic mining techniques lightened some of the backbreaking labour. Water, impounded in reservoirs, was released through pipes in high-pressure jets against the gravel deposits, tearing them apart to wash out the gold. 'Booming' was also practised: sudden release of the dammed-up water so that its onslaught tore out the gravels. These methods were actually destructive of land; eventually, both practices were controlled by law after much terrain and even some towns had been destroyed by their ravages.

All told, California has produced between two and three billion dollars in gold. About two-thirds of this has come from placer mining in the gravels. The other third came from the original gold veins in the mother lode. Usually the original occurrence of this gold is in white or grey quartz veins, which

THIS PAGE: *Agate, a fine-grained quartz, forms from ground water seeping through rock cavities. Its colour variations result from impurities that change as the solutions forming each layer accumulate.* CENTRE, RIGHT: *This specimen is from Rio Grande do Sul, Brazil; all others on this page are from central Mexico.*

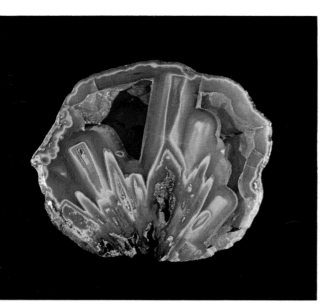

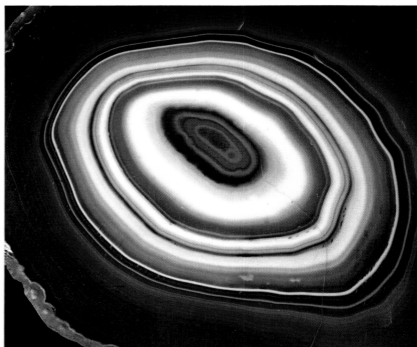

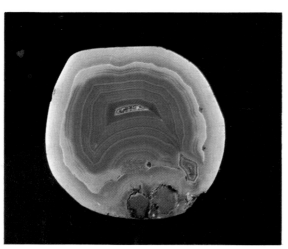

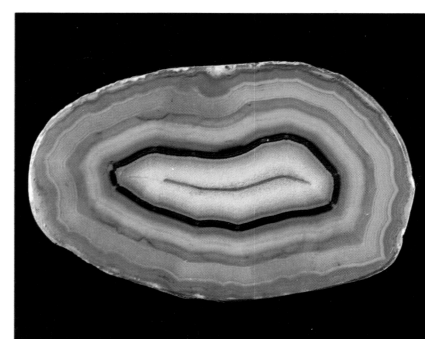

cut through the slates of the Sierra Nevadas in great numbers. Erosion of these veins, especially on the west slopes toward the sea, has been going on for as much as seventy million years, spreading vast accumulations of gold-bearing gravels down the slopes. Some of these gravels were later covered by lava flows. The mother lode extends approximately one hundred twenty miles from near Georgetown to Mariposa.

Gold recovery today, from both placer and hard-rock mining, is considerably more efficient that it was in the early days. Large-scale operations with power machinery and carefully engineered procedures have cut the mining costs sharply by increasing productivity. Better extraction methods have also been developed. The gold can be removed from its crushed ores by the use of liquid mercury, in the amalgam method described in Chapter 4. Another method involves treating the finely crushed ore with solutions of sodium or potassium cyanide. The gold dissolves out of the ore. Zinc metal is added to the resulting gold solution and the gold comes out again free of the waste material in the ore.

It is only through improved methods, reduced costs and the value of other metals sometimes associated with gold that any mines at all continue to operate in the United States. At the onset of World War II, the War Production Board ruled that gold mining was non-essential, and thousands of mines were shut down. Attempting to resume operations after the war, mine operators found that extensive refurbishing of dilapidated machinery and mine workings was necessary. Meanwhile, the cost of labour and materials, transportation, power, taxes, consultant fees, and marketing costs had gone up significantly, while the price of gold was still kept at the 1934 level.

It costs more to mine most of the gold deposits of the United States than the gold itself is worth. Since 1934 the dollar is exactly equal in value to one thirty-fifth of an ounce of gold, by law. It costs more than that in labour, equipment, engineering and other expenses to find the ore, take it from the ground, extract the gold and sell it. The result is that thousands of once-productive gold mines are idle, and ghost mining towns dot the American West.

Meanwhile, as America's cheap gold flows out to the rest of the world, the demand for gold within the United States is rising. As its price remains steady against the rising prices of other metals, its decorative qualities, as well as its good electrical properties, are increasingly attractive to craftsmen and to industry. Drawing on the diminishing gold supply to satisfy these uses drains the American reserves at a rate rapidly approaching one hundred million dollars a year.

What the United States needs is a new discovery of bonanza—from the Spanish, meaning fair weather —ore that can be scooped out of the ground in large quantities.

CORNWALL'S COPPER AND TIN: The discovery by early man that metallic tin could be smelted from the unexciting black mineral cassiterite, tin oxide (SnO_2), is lost in the mists of time — and with it the start of tin mining in Cornwall. The Greek historian Herodotus, writing in about 530 B.C., indicated that the tin trade was already old and that its source was the Tin Islands or Cassiterides. Although Herodotus had failed to locate the Cassiterides, because the Phoenician traders kept their secret so well, there is little doubt despite the disputes of some scholars that the tin came from Cornwall. By the first century B.C. Diodorus Siculus was able to describe the mining operations, and to make it clear that even then veins as well as placer deposits were being

worked. Tin mining has continued in Cornwall to the present day despite periodic fluctuations caused by other mining areas producing more cheaply, and the price of tin is now so high that new mines are being opened. Probably no other mining area in the world has such an enviably long history.

Cornish copper mining has a much shorter history, and before A.D. 1700 was virtually non-existent. Although copper and tin were only rarely produced at the same mines, copper provided the main support of the industry from 1750 to 1850 and spurred the development of deep mining. For just

over a decade from 1850 Cornish mining was in its heyday and was the world's leading producer of both tin and copper. Copper mining then went into a decline, with the mines becoming exhausted below depths of about a thousand feet, and was finished by the end of the century; there are no known reserves.

During Carboniferous times there began the Hercynian mountain-building upheaval of western Europe, in the course of which large masses of granite were formed under the sedimentary rocks of Cornwall. The upheaval produced cracks and fissures in these rocks, many of which provided suitable sites for the deposition of cassiterite and others for sulphides and sulphosalts of iron, copper and lead. A characteristic feature of the Cornish mining region is a zonal distribution of minerals, with tin nearest the granite and other metals further away. Cornwall was submerged by the sea more than once, and the overlying rocks were removed to reveal metalliferous veins at the surface; at the same time the heavy cassiterite was concentrated in placers on the surface and in river beds. Weathering of the sulphide veins produced the secondary oxidized copper minerals which are so highly prized by collectors.

The Cornish ore deposits were rich, but none were really large by modern economic standards. The placer deposits had been worked by tinners from the earliest times, and many parts of the county still show the unmistakable traces of their operations. The veins were usually narrow and nearly vertical, and the mine workings on them were often no wider than was necessary to allow the miners room to work. In the days before machine drills and modern explosives, with primitive machinery to separate ore from barren rock, it would have been hopelessly uneconomical to dig out more rock than was absolutely necessary. Many of the world's mining

TOP: *Early pick-and-shovel miners flocked to California when gold was found at Sutter's Mill.* CENTRE: *James Marshall, who made the discovery.*
FACING, TOP: *A compressed-air hoist in a Cornish mine, from a historic series of underground photos taken by the light of magnesium flash powder.* BELOW: *The Botallack mine, St. Just, Cornwall, no longer standing.*

regions lie in mountainous country, and drainage may often be effected by tunnels or adits driven out to the side of the mountain. Cornwall, by contrast, is relatively wet and low-lying so that deep mining had to wait for the steam pumping engines of Newcomen (1712) and the improved versions of Boulton and Watt (1769) and their successors. The incentive provided by mining to devise and build better pumping machinery did much to speed the growth of the engineering industry, and the ruined houses that once contained the beam engines remain as a characteristic feature of the Cornish landscape. Breaking the hard rock in the mines was once done by 'fire setting', building a fire to heat the rock and then cooling it rapidly with cold water.

147

RIGHT: *The Red Cloud mine in Yuma County, Arizona, produced wulfenite specimens of special interest to collectors.*

BELOW: *Molten sulphur, pumped up by the Frasch process, fills a cooling vat at a sulphur mining plant in Louisiana.*

This slow process was speeded up when German miners introduced gunpowder to Britain; it was first used in Cornwall in 1689, but its use in the awkward and confined spaces of the mine workings caused many serious accidents and fatalities. These accidents were much reduced by a Cornish invention, Bickford's safety fuse, in 1830; safety fuse is still used for firing explosives as the only safe alternative to electrical methods. An excellent account of mining history and the life of the miners is given in Hamilton Jenkin's very readable book *The Cornish Miner* (London, 1948). The skill and hardiness acquired in the mines at home made the Cornish miner a much valued member of mining communities throughout the world, to which he emigrated in times of economic decline in Cornwall. One definition of a mine is 'a hole in the ground with a Cornishman at the bottom of it'.

Because of the large number of mines that have operated at different times in Cornwall, and the diversity of ownership, it is impossible to form a reasonably accurate estimate of the total output of tin and copper. Surprisingly, the value of the copper mined exceeded that of the tin by a wide margin notwithstanding the short time it was in major production. Perhaps the most interesting feature of the Cornish mining area, for our purposes, is not merely the large number of mineral species that have been found there but the high proportion of these that occurred as exceptionally well crystallized display specimens. Among the sulphides and sulphosalts, bournonite from the Herodsfoot mine holds pride of place, but chalcopyrite from St. Agnes, chalcocite from St. Ives and Redruth, bismuthinite from Fowey Consols, and bornite from Redruth are worthy of note. Cornish cassiterite was never found in crystals to match those from Bolivia or Bohemia, but the oxides goethite from the

Restormel mines and cuprite from the Phoenix mine are among the world's finest. Fluorite does not usually come to mind as a Cornish mineral, but the range of form and colour of its crystals exceeds the North of England's best. The almost prismatic calcite crystals from Wheal Wrey are in a class of their own, as are the zoned siderites from Wheal Maudlin. The uranium minerals torbernite from the Basset mines and metatorbernite from the Old Gunnislake mine must attract attention in any collection, and there are other Cornish uranium species still awaiting scientific description. Secondary copper minerals always excite interest, and Cornwall makes up for its lack of the common species malachite and azurite by a startling variety of rarer ones. No listing of Cornish mines and minerals would be complete without mention of Wheal Gorland and its neighbouring mines near St. Day, which many years ago produced such splendid specimens of liroconite, clinoclase, chalcophyllite, and olivenite to make the most exacting connoisseur's mouth water! It is a thousand pities that we are unlikely to find such Cornish treasures again.

THE CITY OF GEMS: Five hundred years before the coming of Christ, invaders swept upon the beautiful island of Ceylon. They drove the native Veddhas, probably direct descendants of local Stone Age men, into the jungles before them. Within two hundred years the Sinhalese invaders had created on Ceylon great cities, vast temples, a wondrous civilization.

Then, in a twinkling, it was all gone. Under new, repeated invasions, which destroyed the temples and cities, the Sinhalese were driven to the southern part of the 25,332-square-mile island. Here in their southern stronghold, these people now mine all the gemstones for which Ceylon is justly famous.

Mining is centred in the districts of Ratnapura, Balangoda, and Rackwana. Ratnapura City (city-of-gems in Sinhalese) sits in a wide hollow, nestling among vegetation-covered hills—a breathtaking tropical setting. The highest part of the island rises sharply from a narrow coastal plain in the south to higher ground, and thence all the way to mountain peaks as high as seventy-five hundred feet.

These mountains are part of the very ancient rocks underlying the island, which have been eroding for hundreds of millions of years—an erosion that has covered much of the rock in the highlands with a layer of reddish clay, making rich soil for the rubber and tea plantations in the hills. Seasonally, heavy tropical rains carry off quantities of this cover in rushing red torrents down deep V-shaped river valleys. Gradually the valleys have built thick layers of sediment; bottom land has been

covered with as much as twenty feet of overburden. The rich red muds of the land around Ratnapura are completely covered by rice paddies, some of which have undoubtedly been cultivated for a thousand years. Under these paddies lie the gemstones of Ceylon.

Mining is very primitive. A miner and his crew lease a spot in a rice paddy to begin operations. All the land here suffers from the complex, highly subdivided ownership, so the miner's location is more likely determined by the lease he is able to obtain than by some good geological reason. A pit is sunk into the clay some ten to twenty feet deep. To hold back the mud, vertical poles are driven against the walls of the pit, and sticks, branches and palm leaves are wedged behind them. Water, however, gets through this, and the pit must be bailed constantly to keep it from flooding. Dis-

carded petrol tins are popular bailing implements. Baskets of clay and cans of water are passed up to the surface and dumped continuously until the bottom of the clay layer is reached. All of this work is so arduous, wet and tropically warm that the men work naked except for loincloths.

Under the clay the iliam, or layer of gem-bearing gravel, is found. Normally it is not very thick, but often it contains rough gem pebbles. All of the gravel is hoisted in bamboo baskets and carried to a nearby stream or water supply to wash off the mud. An expert examiner then carefully hand-sorts the baskets of gravel, and extracts every piece of any potential value. Most of the gravel consists of whitish quartz pebbles, some iron-bearing concretions, and organic matter. However, it also yields an amazing array of gemstones.

There is no greater variety of gems found in any deposit in the world. The sorter is likely to find sapphire of all colours in greatest abundance. There may also be ruby, chrysoberyl, spinel, zircon, topaz, garnet, tourmaline, quartz, moonstone, sphene, cordierite, fibrolite, andalusite, diopside, kornerupine, apatite, sinhalite—a rich variety. Moonstone is the one gem that has actually been found still in the original rocks, others having been found only in the iliam.

The little groups of Sinhalese digging in the clay manage to recover each year a valuable quantity of gems. Because of the difficulty of obtaining sufficient land rights for large-scale earth-moving and drainage operations, mining in this area will undoubtedly remain in this primitive state. We may never know the full extent of the riches that more sophisticated mining methods might reveal beneath the rice paddies of Ceylon.

Copper mining at Chuquicamata, Chile, near the mountainous edge of the arid Atacama Desert. The half-million tons of overburden—waste—being blasted out of the ore body will be carried away to dumping grounds.

Mineral Masterpieces
6

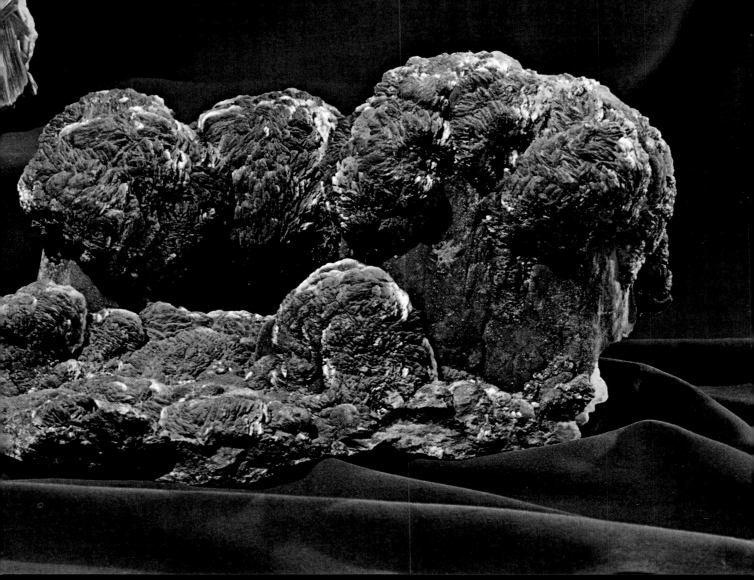

If a masterpiece by Rembrandt or Cellini were reported missing from its honoured place in a great museum, the fact would be headlined in newspapers around the world. Police departments, insurance companies and government offices would all be involved in trying to recover the missing treasure. The theft of important gemstones brings the same reaction, particularly if it is the property of an international personality or one of the world's elite collectors—and, apart from museums, who else can aspire to ownership of such gems? When the celebrated Star of India was stolen recently from the Museum of Natural History in New York, the whole world followed with eager interest the chase and eventual capture of the ingenious thieves. Gems are famous, beautiful, valuable, and generally well protected; evidently this makes the audacious gem theft the kind of caper that captures everyone's imagination.

Unique mineral specimens ought to share the same response. They too are exceptional, valuable, protected, and in certain circles even famous. And yet reaction to the theft of important mineral specimens is usually a quiet affair. When the second-largest uncut diamond crystal in the world on public exhibition was stolen from Harvard University a few years ago, it created hardly a ripple. It was a routine police case of theft. A couple of years later when a large 133-carat uncut diamond crystal on exhibition was stolen from the British Museum, it was even less of a news item. At present, the largest uncut diamond crystal on exhibition is a 254-carat octahedron in the Smithsonian Institution in Washington. It is the only large one left on exhibition, so if it should be stolen its theft might possibly break into print.

The discrepancy in the impact between the loss of a great art treasure and a mineral treasure is strange indeed. Large, well-formed diamond crystals are far rarer than Rembrandt's paintings and just about as irreplaceable. The difference seems to be that mineral museums are relatively poverty-stricken; there is little competition among them for very rare and very expensive items. This keeps prices from being pushed to the heights achieved by paintings and other art works, which are inflated by the competition between wealthy art museums and art collectors.

Most great institutional collections of art throughout the world have their highly prized old masters. If the collection specializes in modern works it will contain art that, in the best judgment of its curators, will become the classics of the future.

Exact parallels can be drawn with the important mineral museums. All the fine old collections have their classic items—outstanding specimens discovered in former years. New mineral museums, beginning without the foundation of an old gift collection heavy in classics, will attempt to acquire the best of the newly mined specimens. And in the future many of these new specimens will become the rare, hard-to-get, and relatively expensive mineral classics.

Very good mineral specimens of any kind are quite rare. There are reasons for this rarity. For one thing, nature just doesn't produce very many extraordinary items. Even when they do exist in a mine or quarry, the obstacles still to be faced before they are carefully preserved in some collection are almost insurmountable. One hundred years ago there was more chance of their being preserved at the mine, since much of the mining was then done by hand, without heavy machinery and powerful explosives. There is nothing more destructive to crystallized specimens than the shock wave of a confined explosion. Also, with the

PRECEDING PAGES: *These newly found specimens meet the connoisseur's standards. Clockwise from far left: scolecite, muscovite, from Brazil; marcasite from Wisconsin; wulfenite from Mexico.*

introduction of heavy, expensive machinery and highly organized mining methods, most companies are concerned that the flow of operations may be disrupted by miners taking time to gather specimens. Some mines control collecting activities by decreeing immediate dismissal for any miner found collecting or selling specimens. This has a definite effect on decreasing the flow of good specimens to collectors and museums—though it never cuts it off completely.

There is a saying among collectors that many thousands more of the good specimens go to the crushers and the mills than ever fall into the hands of collectors and museums. Some mining companies have helped improve the situation by permitting collecting by miners outside working hours as a sort of fringe benefit. Others have permitted retired miners to bolster their incomes through the collection and sale of specimens. Whatever the means, there is a continual trickle of specimens from the mines.

Once out of the mine and into the hands of a collector, however, there is still no guarantee that the specimen will survive. When either the miner or the collector transports the specimen from an inaccessible mine to civilization, there is the ever-present possibility that it will suffer irreparable damage, thus perishing as an important specimen. Having arrived at the collection, there is again no guarantee of its survival. Through ignorance of proper methods for trimming or cleaning, or through carelessness in handling, the specimen can be lost in an instant. Even supposing that it survives and is carefully guarded during the collector's lifetime, it may face eventual destruction. His heirs, through neglect, lack of interest, or mishandling during packing, can easily decrease or destroy its value as a collector's item.

Assuming that our specimen is tough enough to have survived all this, and is deposited with a collec-

The Imperial Easter Egg, made in 1897 by Fabergé, the Russian Court jeweller, and presented by Tsar Nicholas II to the Empress, is a masterpiece in gold studded with precious stones. Exceptional old mineral specimens, particularly from now-exhausted sources, are similarly unique and unobtainable.

tion, it may face other hardships. Libraries, local schools, historical societies and universities have been recipients of collections. By and large, the care given them is minimal, and innumerable collections that were obviously important at one time are now practically worthless. The chances of survival, then, from the mine to a secure final resting place in a well-cared-for and stable museum collection, are slim indeed.

But among the surviving specimens will be a few that are the finest of their species. The more outstanding the specimen, the better its chances, for it is likely through the years to get consistently good care. Some will be the largest crystals of their kind. Others may have the most unusual colour for their species. Still others become important because the mineral deposit they came from is now exhausted and no similar deposits have been found.

The best places to see the old classics are at large museums that have been in the mineral-collecting business for many years. The mineral museums of the United States are relatively young compared to those of Europe. However, some have assumed classical status, like American art museums, through purchases of old collections and through valuable gifts. Many fine modern collections will not drift to museums for perhaps another fifty years or so. Usually, therefore, mineral museums are blessed with old classics but lack a sampling of new ones. This is a typical picture of most mineral museums in the world, except for a very small number that have won a share of new discoveries by developing aggressive soliciting and purchase programmes.

Collectors and museums concentrating on the acquisition of fine, newly discovered specimen material are in the same position as those collecting contemporary art. Contemporary art is relatively inexpensive and a bit of a gamble. No contemporary artist is likely to get a hundred thousand for one painting. Similarly, contemporary mineral specimens are less expensive than old ones, but they are a less speculative investment than modern art, since their beauty, quality, rarity and general desirability are more obvious. Important new finds are recognized at once. They are—or will be—the new classics.

It would be a mistake, for instance, for the collector to avoid buying superb specimens of calcite from Mexico because they are at present common and cost only between one and ten American dollars. Their future will very likely follow the precedent set by the famous calcite specimens from England that were so common twenty-five or thirty years ago, which sold at that time for similar low prices. Now, quality English calcite specimens are just about unavailable at any price.

What are these mineral 'Rembrandts'? And how are they recognized? It is a simple matter to buy a colour-illustrated catalogue of the paintings of almost any great artist, complete with biographical information. Such guides do not exist for mineral specimens. Collectors and museum curators, through comparison of collections and continual contact, have passed information along to each other, but only in bits and pieces. The museums themselves are the only catalogues.

Visitors to museums can usually be sure that the finest specimens in the collection will be displayed and that among them will be many classics. To the alert visitor, making comparisons from museum to museum, certain repetitions become evident. Each collection will probably have fat, light blue crystals of topaz from Russia; long, splendent, metallic shafts of stibnite from Japan; and black, opaque, shiny tourmaline groups from Pierrepont, New York. These are some of the classics.

LEFT: *Tourmaline from Elba, Italy.* TOP, RIGHT: *Stibnite, Ichinokawa, Japan.* LEFT: *Realgar, Transylvania, Rumania.*

157

The difficulty lies in retaining exact mental pictures of these specimens while comparing various collections. Also, as mentioned before, most museums possess only old occurrences. This is one potential pitfall for the collector. His idea of what is classic and important may become distorted. He may hunger after the unobtainable old things in the museums while ignoring the new. The new specimens seem too inexpensive to be important, they are too plentiful, everyone has them and, worst of all, they have not been immortalized by descriptions in mineral textbooks or by the occasional publication of photographs in mineralogical literature. Some of them are described below and pictured throughout this book to encourage the collector. There is good reason to anticipate that one day these

presently available specimens will be 'Rembrandts' along with the old classics.

Appreciation of the classics by the general museum visitor and mineral collector depends chiefly on three things. First, there must be a sensitivity to beauty, since for many of the pieces this is the major claim to fame. There must also be an awareness of the natural history of mineral specimens. For example, the rough, blocky crystals of andalusite from Delaware County, Pennsylvania, are hardly objects of beauty, but they are some of the largest and best-formed groups ever found. Finally, an ability to compare specimens and to discriminate among them is indispensable. To the uninitiated, the specimens in a mineral exhibit are all interesting or all beautiful or all the same—at first. They wonder what the experienced collector sees as he moves slowly through the exhibition, carefully examining each piece. Of course, he is making mental comparisons between these specimens and others he has seen. He is also memorizing details—he may actually take notes—of old classics, which will teach him how to recognize and search out new classics for his own collection.

As an indication of what the collector may want on his 'desiderata' list, a very small sampling is given here, listed in the chemical groupings featured in most modern systems of mineralogy. Famous old classics are included, along with some sparkling newcomers. A complete description of classics from even one group could be a volume in itself. Almost every specimen pictured in this book might well be included among the new and old classics.

NATIVE ELEMENTS: Native elements are those that are found free in nature. Both metallic and non-metallic elements among the minerals are well represented by superb specimens. Of the non-metals, *sulphur* is the most attractive. The best

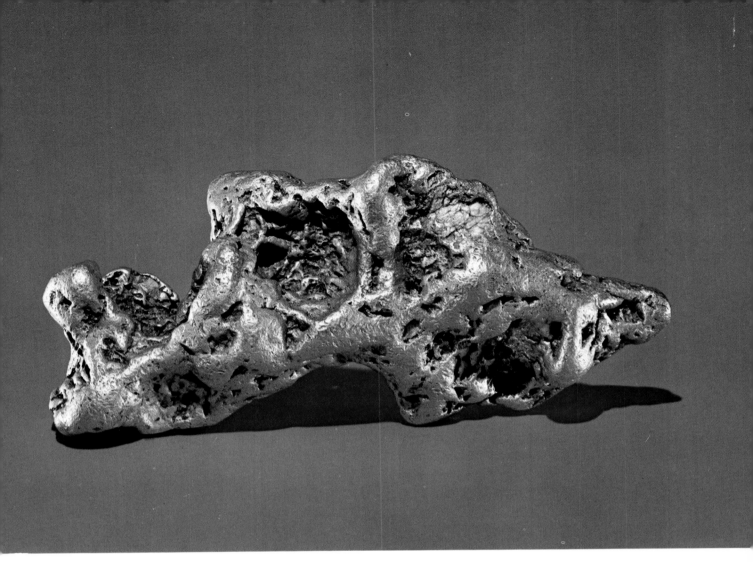

examples, still available, come from the mines at Agrigento, Cianciana and elsewhere in Sicily. Bright, clear yellow crystals from microscopic size to four or five inches across occur in homogeneous groups, or associated with outstanding, shiny, white, chisel-pointed crystals of celestine or large, hexagonal, barrel-shaped crystals of aragonite. Some of the new specimens, from San Felipe in Baja California, are as bright, colourful and well formed as Sicilian sulphur, even though the crystals do not exceed half an inch.

The best specimens of *diamond*, another non-metal, have come from South Africa. It is still possible to obtain, at a good price, sharp, transparent octahedrons as loose, single crystals of various sizes. However, good-quality crystals that measure half an inch or larger are almost prohibitive in price for most collectors. Because of newer mining methods, using powerful explosives and large earth-moving machinery, it is often impossible to obtain crystals still embedded in rock. In fact in most of the old specimens of this kind found in museums, the crystals have been glued in, although both diamond crystal and rock are genuine. Well-shaped, tan diamond crystals in various forms, measuring three-eighths of an inch or more, but not of gem quality, have been obtained from the Congo.

Silver specimens in sheets, wires and very sharp crystals occur in calcite in the rich ores at Kongsberg, Norway. The beautifully twisted, sculptured, thick wires can be as much as a foot long and as thick as a

FACING: *Brown siderite rhombs make quartz from Allevard, France, a classic item.* ABOVE: *Large gold nuggets rarely persist intact. One of the largest known is this eighty-two ounce specimen from a California placer mine.*

man's wrist. From the copper mines on the Keeweenaw Peninsula in Michigan the best silver
specimens recovered tend to be branching, heavy,
fern-like growths of fairly distinct crystals associated
with copper. The *copper* specimens from these mines
are the finest in the world. They can be obtained in
almost any size, all of high quality. Some are
aggregates of crystals, one inch or larger, clustered
like walnuts in branching groups. Others may be
delicate branching fans of coppery fern fronds.
Remarkably good copper specimens have appeared
in recent years from the New Cornelia pit at Ajo,
Arizona. These, too, have been found in small and
large aggregates of one-inch crystals and as branching
tree-like growths, all of bright coppery colour.

Gold is found in many places and in large specimens. Perhaps the most attractive have been found in California and in Rumania. From the mother-lode country of California have come the finest crystallized specimens. The specimen of flat clustered crystals in the Cranbrook Institute, from the Red Ledge mine in California, is perhaps the finest gold specimen in the world. Many large nuggets of attractive form have been found also in placer deposits in California. Specimens from Rumania tend to be lighter in colour because of the presence of some silver.

Platinum, another noble metal, is rarely found anywhere in its original rock deposits. Extraordinary nuggets, some weighing many ounces, have been found in gravel. The very large rounded nuggets occurring in the headwaters of the Tura River in Perm Province in Russia are the best known. Similar nuggets, almost indistinguishable from their Russian counterparts, come from the Department of Cauca in Colombia.

A few years ago the mineral *allemontite* might have been considered scientifically interesting, but hardly a very desirable show species. The original material, from Allemont in France, was dull and grey. However, this natural arsenic-antimony alloy has since been found at Moctezuma in Sonora, Mexico. The Sonora specimens are bright, shiny, rounded, colloform masses, which make very attractive additions to a collection.

SULPHIDES: Many of the mineral ores from which commercial metals are extracted are classified chemically with the sulphides. As expected, then, mining activities have turned up extraordinarily good specimens of several sulphide minerals. The silver sulphide, *argentite*, is not a beautiful mineral, but like all silver minerals, is rather rare. When well crystallized, it becomes a collector's item. Flattened

and malformed and sometimes skeletal crystals and groups of crystals from Arispe in Sonora, Mexico, are highly desirable. Parallel and stacked crystal dendrites, as well as crystals up to an inch or more, from the Freiberg district in Saxony, Germany, are also among the best.

The copper sulphide, *chalcocite*, is best known from two occurrences. In the United States, fine groups of crystals, tarnished black, were found near Bristol, Connecticut; some are still circulating in private collections. More, and perhaps better, specimens have come from Redruth and St. Just in Cornwall, England. These occurred as hexagonal-looking black plates, often stacked on each other in piles, with some of the crystals as much as an inch in diameter.

One of the commonest lead sulphides, found in many places in excellent specimens, is *galena*. The lead-zinc mines in the Tri-State area of the United States—which includes the corners of Missouri, Kansas and Oklahoma—produce good speci-

FACING, TOP: *Benvenuto Cellini's salt cellar, a well-known Renaissance masterpiece in gold.*
LEFT: *Mexican tourmaline.* ABOVE: *Buergerite—found only in San Luis Potosí, Mexico.*

mens by the ton. Most of them are in bright, shiny, well-formed cubes, but some are malformed, elongated or skeletal. Almost any form or variation possible with galena crystals can be found among these specimens. In some of the most attractive, the galena cubes are dusted with a sparkling coat of tiny golden marcasite and pyrite crystals.

The crowning glory of the sulphide division of a collection will be *stibnite* from Japan. This is an antimony sulphide formerly found at Saijo in Iyo Province, on the island of Shikoku. Crystals in these groups are up to two feet long and some are three inches across. Many are a shiny, steely grey, and are very well formed. They are so magnificent compared to even fine stibnite specimens from other places that they are instantly recognizable.

Pyrrhotite, an iron sulphide, is still available in classic specimens from at least two occurrences. The mines at Trepca, Yugoslavia, have been producing good specimens for years. Many of the old classic pyrrhotite specimens came from Mexico. These were not available for many years, but new

ones are beginning to appear from mines in Chihuahua. Trepca has supplied hexagonal-looking crystal plates up to four inches across, which are usually stacked or arranged in rosettes. The Mexican crystals tend to be more sharply formed and thicker, although not so large in diameter. Both in Yugoslavia and in Mexico, they occur with an attractive bronze tarnish. Magnificent crystals formerly came from the Morro Velho mines in Brazil.

SULPHOSALTS: The sulphosalts normally occur with sulphides in ore deposits, and are just a little more complex than sulphides in composition, having some antimony or arsenic along with the usual sulphur content. The silver sulphosalts are among the most coveted. Of these, *proustite*—called 'ruby silver' by collectors for its red colour—is the most attractive. At its best occurrence, in the Dolores mine at Chanarcillo, Chile, it has been found in bright red single crystals, two to three inches long. It also occurs as larger dendritic groups of crystals. An extraordinarily large mass of proustite crystals is in the collection at the Harvard Mineralogical

Museum and the British Museum has a transparent crystal over three inches long and nearly two inches across. Sometimes proustite is so clear that it can be faceted as gemstones, even though the stones are much too soft for any practical use. Ruby silver specimens command prices many times the value of the silver that they contain.

Specimens of *tetrahedrite* are easily recognized by the collector, because they are often sharply crystallized in the tetrahedron crystal form from which the species gets its name. The most attractive of these are crystals from Clausthal in the Harz region of Germany, and from the Herodsfoot mine, Cornwall, England. Crystals up to one inch on an edge are completely coated with a gilding of chalcopyrite, which emphasizes their form and gives colour to the blackish crystals. Small but bright and sharply developed crystals associated with pyrite and bournonite on white calcite have been coming in recent times from the Concepción del Oro area in Zacatecas, Mexico. The Herodsfoot mine, in the 1860s, also produced unequalled specimens of 'cogwheel' bournonite, a sulphantimonide of copper and lead; these are greatly sought after by collectors and are almost impossible to obtain.

HALIDES: The halides are minerals containing the elements fluorine, chlorine, bromine or iodine in their compositions. Two of them, *halite* and *fluorite*, are very common. Halite, sodium chloride, is salt. Salt crystals, usually cubes, can be collected almost everywhere—even in a salt cellar, in which every grain under magnification appears as a cube. In the enormous salt mines of Galicia and Wieliczka in Poland, many crystal groups are found with some cubes ten to twelve inches on an edge. There is a magnificent group, standing three or four feet high, in the Vienna Natural History Museum, with glass-clear crystals of this size. Handsome crystal groups

are coming from many places in the United States, one being the salt lake-bed deposits of Searles Lake on the San Bernardino desert in California.

Fluorite, calcium fluoride, also crystallizes well in many of its deposits. The quantities of very beautiful specimens that once came from the English lead mines of Weardale and Durham, and the mines at Alston Moor, Cleator Moor and elsewhere, are no longer being produced. So many were originally circulated, however, that it is still possible to obtain them from dealers. The perfection of their crystals and the wide range of colour are difficult to match. To replace them, the mines near Rosiclare and Cave-in-Rock in southern Illinois, U.S.A., are now injecting into private collections large quantities of excellent specimens in a range of colours from tan to purple to sky blue. They lack the clarity and the strong blue fluorescence of English fluorite specimens, but are still very attractive.

OXIDES: So many oxide minerals occur in attention-demanding specimens with excellent crystals that selection of a small group is difficult. Certainly

Copper minerals in three widely different forms. FACING, FAR LEFT: *Dioptase, copper silicate, from Guchab, South West Africa.* LEFT: *Chalcopyrite, copper iron sulphide, from Pennsylvania.* ABOVE: *Malachite, copper carbonate, from Bisbee, Arizona.*

among them would be the common iron oxide *hematite*. Well-crystallized examples of hematite are black, and so splendent as to appear polished. Sharp, curved crystals from Elba are often arranged in roughly parallel, stacked groups that may have a brilliant lustre or peacock coloured tarnish. From Brazil and Switzerland come specimens, also brilliant, on which the flat crystal plates are precisely arranged in unfolded rosettes. Those from Brazil tend to be heavier and more masculine-looking, while those from Switzerland are delicate and feminine, perched lightly on sparkling crystals of quartz and feldspar. In recent years there has been a trickle of shiny, blocky crystal groups coming from Bahia, Brazil. And any listing of hematite classics would have to begin with the shiny, black, rounded 'kidney ore' specimens from Cumberland, England.

A unique classic occurrence for the iron and manganese oxide, *franklinite*, should be mentioned even though specimens are seldom available. It is a rare mineral except at Franklin, New Jersey. There it has been found in well-formed, black octahedrons of various sizes, often embedded in white marble. The largest of these, measuring seven inches on the edge of the octahedron, is in the Smithsonian Institution.

CARBONATES: *Witherite* and *aragonite* are chosen to represent the classic carbonates because it is possible to acquire them now in excellent specimens. In the past, the most famous examples of witherite were in white, hexagonal-looking, pyramid-shaped crystals from Northumberland, England. These are no longer available, except from old collections. Meanwhile, the fluorite mines of southern Illinois, U.S.A., have produced many well-formed crystals of equal or larger size, up to three inches across. They are pale tan rather than white, and are flat on the ends rather than pointed, which

gives them a more squat, but still attractive appearance.

Aragonite is found in many places and in many forms. Perhaps the best of the currently recovered aragonite classics come from Sicily and Spain. Pieces from both these places are identical with the specimens that came into collections fifty or more years ago. The Sicilian crystals are white and brilliant, occurring as hexagonal-looking twins on yellow sulphur. Most of the Spanish crystals are the same kind of twins and are just as sharply developed and brilliant. They differ from the Sicilian in that they usually occur as isolated crystal twins, rather than in groups, and are reddish-lavender to brown in colour.

SILICATES: Enough good *tourmaline* specimens have come in the past from Elba, Madagascar, Brazil, New York, Connecticut and elsewhere to build an entire classic collection of this mineral alone. Even with the excellent supply, however, good tourmaline specimens have never been inexpensive. Almost all of the material cut into tourmaline gemstones comes nowadays from the state of Minas Gerais in Brazil. Often a crystal will escape cutting or survive because it is of less than cutting quality. As a result, some are always available for collections.

Other mineral-specimen tourmaline crystals, in soft tones and combinations of green and pink, are being recovered from the pegmatite mines of San Diego County, California. Some of these are six inches or more in length and more than an inch in diameter, and some are completely developed on both ends.

Near Santa Cruz in Sonora, Mexico, black, lustrous, highly modified crystals of tourmaline are being found in the tungsten ore deposits. They so much resemble the shiny black crystals found

half to three-quarters of an inch thick, and usually opaque, though a few were transparent. Additional transparent, yellowish, rounded crystal masses suitable for gemstones were found in the gravels of Mogok, Burma. The newer specimens, coming out in a flood from Charcas in San Luis Potosí, Mexico, are far handsomer. They occur in groups of chisel-shaped crystals up to about four inches long. Although simple in form, they are bright, colourless to very pale pink, and transparent at the tips. Often they are in attractive groups jutting out like blades from the matrix material, and are sometimes associated with calcite crystals.

The list of classics, old and new, could be extended to include hundreds of others. The old *datolite* specimens from Westfield, Massachusetts, have given way to new ones from Prospect Park, New Jersey, and Loudon County, Virginia. Old *sphene* specimens from Switzerland are joined by excellent new specimens from Brazil. Famous *epidote* groups from Austria and Alaska are beginning to share honours with the new ones from Baja California. The shiny, black, *ilvaite* crystals from Idaho are even better than any recovered from Greece or Elba in the past.

If the connoisseur and collector now studying the old classics keeps an open mind, he soon finds that he is living in a time of the discovery of new mineral 'Rembrandts'. Any or all of the specimens merit inclusion in his collection.

years ago at Pierrepont, New York, that they are often confused with them. In the same deposits there are also strange groups of black tourmaline, looking like tapered plumes, with deceptive velvety-looking surfaces.

Mexico is also producing specimens of *danburite*, another silicate, to replace the fine classic crystals that came at one time from the Toroku Mine on Kyushu, Japan. The Japanese crystals were single, well-developed shafts about four or five inches long,

Shapely masses of entangled silver wires are typical of the fine specimens of native silver from the rich ores at Kongsberg, Norway. The fantastically twisted wires are sometimes a foot long and as thick as a man's wrist.

The Science and Its Framework

7

The ultimate goal of the science of mineralogy is to know as much as can be learned about mineral species and their relationships, both to each other and to the environments in which they form. Sometimes the knowledge can be put to immediate practical use but often, as is true in all sciences, it is sought for its own sake.

Mineralogy is one of a large number of working divisions of the earth sciences. It forms a bridge between geology on the one hand and chemistry and physics on the other. It is more closely related to geology than to chemistry and physics because it is concerned primarily with materials of the mineral kingdom. It is helpful, then, to understand the geological framework in which the science of mineralogy operates.

Geology is the study of the earth. Geological sciences include studies of rocks, ores, minerals, mineral fuels, fossils and the natural processes from which they are all derived. These earth sciences, being broad in their coverage, rely heavily on other sciences. Chemistry, physics, mathematics, engineering, zoology and botany are all involved in one way or another; their special contributions are vital in unravelling the story of the earth.

In recent years geology has expanded to take the planets under its wing. The United States Geological Survey is now preparing accurate maps of the moon and in several locations geological laboratories are being readied to study the first samples of minerals brought back from the moon.

THE GEOLOGICAL SCIENCES: The very broad spectrum of geological activities is reflected in the branches of the science that are generally recognized.

Geochemistry is the study of earth materials that seeks information about the chemical conditions that caused their occurrence, their movement in the earth and their present distribution in the earth's crust. Increased understanding of nature's chemical processes also has high practical value in helping to predict the occurrence of ore deposits.

Geologic Mapping is a method of making an inventory of the contents and placement of various sections of the earth's crust for later research or economic exploitation. Photogeologic mapping is a specialized method of mapping based on the use of precisely located aerial photographs.

Geomorphology is concerned with the contours, forms and details of the exposed parts of the solid earth and the natural processes that caused them. Lakes, mountains, valleys, plains and beaches are in the province of the geomorphologist.

Geophysics is a complex branch of the science, comprising: seismology, a study of earthquakes and tremors; hydrology, a study of the occurrences and behaviour of water in the crust of the earth; oceanography, a study of the oceans, their movements, currents, temperatures, composition, and the nature of ocean bottoms; terrestrial magnetism, a study of variations in the earth's magnetic field and the reasons for them.

Glaciology is the detailed study of present and long-past glacier activity. From it we derive information about the sequence of changes in earth temperatures, the characteristics of earth's oceans, and the effect of glacial periods on other geological processes and on life.

Historical Geology progresses toward the unravelling of the past history of the earth, based on evidence permanently preserved in rocks and fossils. Historical geology has been highly successful in reconstructing the geological story of the earth. The basic outline of the story is supported by sufficient evidence so as to be beyond doubt, although the details are constantly subject to modification as new evidence appears.

PRECEDING PAGES: *A sample of olivine-rich dunite from North Carolina, ground to fraction of a millimetre thickness, is magnified five hundred times and photographed through a petrographic microscope.*

Micropaleontology is the study of microscopic fossils. The remains of very tiny animals called foraminifera, for example, have long been indicators of the age and composition of sedimentary rocks. This information can be especially useful for locating valuable petroleum deposits. There have even been attempts to set up a separate science of palinology, a study of fossil pollen.

Mineralogy, is the branch of geology that provides information on the classification, composition, internal structure, and chemical and physical behaviour of minerals, the materials with which all of geology is ultimately concerned.

Paleobotany is fossil-plant study, in which the skills of geologist and botanist combine to analyze the occurrences and evolution of ancient plant types.

Paleontology, an all-inclusive science, gathers information concerning all plants and animals that have ever existed on the earth, and is invaluable to historical geology. Paleontologists maintain that 'unless the remains of an animal still smell, they are the study material of paleontology'. Scientists in this field often join the skills and knowledge of biology with those of geology.

Petrography provides methods for the systematic description and classification of rocks. Petrography requires a thorough familiarity with mineralogy so that minerals can be recognized as they occur in rocks. Accurate rock descriptions are basic to any study of rock formation and change.

Petrology is the study of the natural history of rocks, their origins, present condition, alteration and decay. The science relies heavily on mineralogy, but deals with the assemblages of minerals composing rocks rather than with individual minerals. These composite samples are much more revealing of processes within the earth than are the individual minerals.

Ferns, cycads, lepidodendrons and other foliage, in an artist's conception of a prehistoric carboniferous forest. Coal was produced as thick layers of vegetation were broken down by time and pressure, leaving fairly pure carbon.

Sedimentology is a specialized geological field, because the problems in understanding rock sediments are different from those posed by other kinds of rocks. The original rocks from which the sediments came usually leave little evidence of what they were before erosion tore them apart. A special procedure has been developed for the study of sedimentary rocks and the processes that formed them, to overcome the difficulties resulting from the fragmentation of the original material.

Stratigraphy uses evidence from several other branches of geology to describe the age, character, thickness, and sequence in which rock layers formed.

Structural Geology is the field of the geologist interested in the folding and breaking of rock layers, resulting in the deformation of the earth's crust. Studies of mountain building are typical of these researches.

THE MODERN GEOLOGIST: Our civilization depends on plentiful supplies of minerals and mineral fuels. These resources must be searched out and developed before they can be converted to practical use. An increasing number of highly specialized geologists supply the skills to do this. There are exploration geophysicists, petroleum geologists, economic geologists, ground-water geologists, engineering geologists, and even military geologists. Thousands of colleges and universities throughout the world now offer degrees of various kinds in the earth sciences.

It must already be obvious that the stereotype of a geologist as a grizzled prospector, wandering in the rocky wilderness with his donkey and his rock hammer, dates back to the days of silent films. A geologist may spend his entire career working in the field on various projects, but it is equally possible that he will choose to confine himself to full-time laboratory research work.

In the field the geologist, working with modern equipment, carefully measures and traces rock outcroppings to provide accurate observations from which geological maps can be made. These maps will eventually show details of hills, valleys, streams and other surface features of the countryside. They may also show the nature of the underlying rock deposits that formed the surface features, give information as to their origin, or even indicate rock deposits for potential economic development. The geologist will also collect samples for later detailed laboratory study.

An office-bound geologist assembles and correlates information obtained through field parties and other kinds of research, and attempts to extract useful meaning from the information.

In the laboratory, the earth scientist carries out physical and chemical studies of the materials of the earth, assembling more information and testing various theories about their characteristics and origins.

There are a number of professional geological endeavours in which many earth scientists have been involved, each contributing his part. One of the most fascinating examples of such projects has involved hundreds of geologists from everywhere in the world for many years. They have slowly assembled, as in a jigsaw puzzle, the geological history of the earth.

Shortly after the Reformation, about the time other physical sciences were getting their start,

Archbishop Ussher of Ireland made a name for himself by working out, from the Scriptures, a date for the creation of the earth. He set it at 9 A.M. on the morning of 26 October 4004 B.C. Probably because of his exalted position, the date was somehow entered as a footnote in the King James Bible.

For decades the true history of the development of the earth's features was obscured by the popular prejudice created by this reckoning. Since, according to the Archbishop, the existing features of the earth had been formed in only fifty-six hundred years, it was obvious that everything must have been done in a terrible hurry. All the gravels and unconsolidated sediments were thought to be remnants of Noah's great flood. Mountains

FACING, TOP: *Fossil trilobite in shale, found near Lebanon, Pennsylvania.* BELOW: *Fossil ammonite from Dorset, England.* TOP: *In 1833, an artist's conception of a prehistoric scene mixed flora and fauna from many geologic periods.*

were thought to have been thrown up in an instant, with great violence. Deep valleys were considered the result of clefts left by violent earthquakes that accompanied these rapid changes. There just had not been enough time, since the Creation, for any other earth-shaping process.

One hundred and thirty years later a Scottish geologist, James Hutton, made the first serious attack on these notions. In 1785 he advanced his Law of Uniformitarianism, which stated that the geological processes taking place now are the same ones that have always been taking place, and that the existing features of the earth were formed in the same way in which similar features are being formed today. If this could be substantiated (as it later was), it would indicate that the earth required enormous amounts of time to form its features. The earth, instead of being a few thousand years old, was a few billion years old. Every bit of geological, chemical, astronomical and physical evidence now supports this thesis, but geological proofs were long and slow in coming.

The scientific study of the earth's actual age was given impetus with the development of two ideas.

One, the self-evident Law of Superposition, holds that in any series of sedimentary rock layers the most recent sediments, or the younger rocks formed from the sediments, are on top, and the older at the bottom. The other, introduced by geologist William Smith in England before 1800, is the concept of index or guide fossils. Smith found that in any given rock layer only certain kinds of plant and/or animal fossils were preserved. He did not know why they differed in various rock layers, but it was obvious that each layer had its distinct fossil trademarks.

The evidence for both the Law of Superposition and the significance of index fossils is somewhat affected by the erratic behaviour of the earth's crust. Some areas are so disturbed by the folding of rock layers that older rocks have been pushed up on top. In other places the layers are interrupted or duplicated by breaking and shifting motions.

Yet there remains overwhelming geological evidence to support these important ideas. Geologists, studying rock layers around the world, have been able to assemble the layers in a continuous series called the geologic column. Its contents record a sort of time chart of the sequence of geologic events

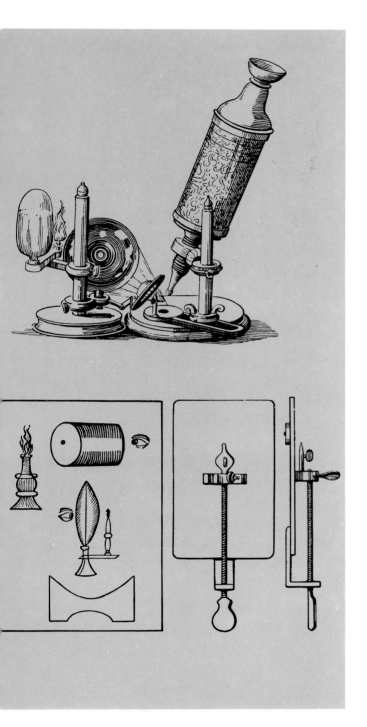

hundred miles thick if piled all in one place.

After a hundred and fifty years at the work of dating rock layers—or stratigraphy—geologists have become quite skilled at it. Following are some of the tests they have developed which, taken together, provide reliable information.

Lithologic similarity: When two beds of sedimentary rock are found, separated by considerable distance but made of identical rock materials, they are very likely parts of the same layer.

Continuous tracing: Any rock layer that can be traced continuously from one place to another through exposures and outcroppings is obviously the same age throughout.

Similar thickness: Rock layers in two different sequences may very well be of the same age if they are identical in thickness.

Similar positions: If two dissimilar layers of rock in two localities occupy the same position in a sequence in which other layers are known to be identical, then they too are the same age.

Gradation over a distance: A single rock layer traced over a distance may be seen to blend into a different layer with considerable interpenetration at their edges. The two layers are the same age.

Identical fauna: Comparing the fossils in two widely separated rock layers will show whether or not they are at the same evolutionary stage. If they are, the rock layers are the same age.

Dissimilar fossils: Two rock layers in isolated places may contain different sets of fossils but still be the same age. If both sets of fossils occur elsewhere in one rock layer, then they are known to represent the same age. This means the original two rock layers are also the same age.

Using these guides, and developing the geologic column, a very detailed and lengthy history has slowly come to light. For convenience, the historical

that has taken place on the surface of the earth. No one part of the earth has a complete section of the column, but in spite of this it has been possible to piece together a series of layers that would be one

FACING, FAR LEFT: *Archbishop James Ussher arbitrarily dated the Creation at 4004 B.C.* LEFT: *Geologist James Hutton and chemist Joseph Black made contributions to eighteenth-century science.* ABOVE: *Early microscopes.*

173

geologist divides this history into eras which begin and end with great crustal revolutions. These revolutions interrupt the steady erosional and sedimentary pattern, which goes along undisturbed for millions of years between upheavals. The most recent revolution is the one that, through tremendous uplifts, gave birth to the Alps, the Himalayas, the Andes, the Sierra Nevadas and other mountains. These revolutions divide into five parts the time during which there has been life on earth.

According to stratigraphic dating, first sugges- tions of fossil life appear in rocks over five hundred million years old. Traces of certain low plant forms called algae, and some primitive sponges, have been found in these rocks. Evolution continued, and in the *Paleozoic Era*, beginning five hundred million years ago, fish and some conifers—cone-bearing trees— began to develop. By two hundred million years ago, the dinosaurs had started developing in the *Mesozoic Era*. Towards its end, seventy million years ago, they went into decline. Mammals and flowering plants then made their ascent in the *Cenozoic Era*.

Man has put in his appearance among the mammals only in the last million years or so. The anthropologist D. J. Mahoney in a 1943 review of the antiquity of man, portrayed the development graphically when he said, 'We may imagine ourselves walking down the avenue of time into the past and covering a thousand years at each pace. The first step takes us back to William the Conqueror, the second to the beginning of the Christian era, the third to Helen of Troy, the fourth to Abraham, and the seventh to the earliest traditional history of Babylon and Egypt. . . . One hundred and thirty paces takes us to Heidelberg man, and about a quarter-mile to the oldest undoubted stone implements of Europe. And should we decide to continue our journey until we meet the most ancient fossil organisms, the journey would exceed two hundred and fifty miles.'

The task of cooperatively filling in all the details of the journey is one of the greatest monuments the science of geology has erected to itself. In this work, the science of mineralogy makes constant and vital contributions.

MINERALOGY, THE SCIENCE: For convenience, the science of mineralogy has traditionally been divided into crystallography, physical mineralogy, chemical mineralogy and descriptive mineralogy. A mineralogist's work may be in any or all of these divisions, particularly since they are arbitrary and may overlap considerably.

Crystallography had its beginnings in the measurement and study of relatively large and well-formed crystals. Now it has become a major science of basic importance in chemistry, physics, metallurgy and biology.

Previously its major concern was with the external shapes and symmetries of crystals. These were accurately measured by the use of a narrow beam of light, reflecting from crystal faces. Now almost total interest has turned to their internal structures, as revealed by X-ray techniques. External symmetry is considered merely the necessary effect of internal structural causes.

Chemical mineralogy concerns itself with the elements of which minerals are made. Compositions and variations in composition are determined by traditional wet chemistry methods, using test tubes, chemical solutions, acids and so forth, or by electronic intruments. This information is later compared with information about the elements' structures to deter-

FACING: *Alexander the Great's diving boat was probably a legend, but man has always made efforts at undersea exploration.* ABOVE: *The U.S. Navy's submarine Trieste, and other advanced devices, provide data never before available.*

mine the effect they have on each other. Chemical mineralogy is also interested in the association of various species as they occur together in nature. This gives some indication of the chemical environment in which they grew. In turn, this reveals to the chemist the kinds of chemical activity that nature permits in mineral deposits, and how the natural activity compares with chemical reactions occurring in the laboratory.

Physical mineralogy is concerned with the physical characteristics of mineral species. Their behaviour, hardness, tenacity and other physical characteristics show a close relationship to their compositions and structures. Because many of the physical properties determined for minerals have proved to be significant in industry, they are being increasingly investigated.

Descriptive mineralogy attempts to assemble enough information about each mineral species so that it can be sharply defined and not confused with any other. This usually means, in modern times, a chemical analysis to fix its composition, and a study of its crystals to determine their shape and symmetry, as well as X-ray analysis to find the symmetry of the structure and the size and shape of the unit cell. An optical analysis determines its refractive indices and other behaviour toward light.

The scientific language used for describing and defining a mineral species is exceedingly abbreviated. Yet, when translated, mineralogical language conveys all the significant facts. A remarkable amount of time, energy, and sometimes complex equipment is involved in fashioning the description. As an example, we will translate and investigate the formation of an abbreviated description of the mineral

species wurtzite.

Crystallography—Hexagonal-P; dihexagonal pyramidal-6mm a : c = 1 : 1.6349.

This means that the mineral crystallizes in the hexagonal system; information is given about the mineral's symmetry; the ratio, a : c, explains that the vertical axis of a wurtzite crystal is 1.6349 times as long as any one of the three horizontal axes. The crystallography may be determined from measurement of well-formed crystals, or from X-ray examination. It is best if both methods can be used as a check on each other.

If crystals are measured, the instrument used is a two-circle goniometer. This instrument was perfected in the late 1800s in Austria by Professor Victor Goldschmidt and his mechanic, who gave his name—Stoe—to several fine old instruments that are still giving good service.

The crystal to be measured is mounted on the goniometer, which is ingeniously designed so that the crystal can be turned in any direction. A narrow beam of light, coming from a light tube, is bounced off one crystal face after another, into a telescope sight at the eye of the observer. The position of the crystal face is recorded from a reading of numbers marked on two movable circles on the instrument. Each set of numbers is then plotted as a kind of graph. From this graph, called a gnomonic projection, the symmetry of the crystal can be read. The graph can also be used to make a drawing of the original crystal by converting its measurements according to standard drafting procedures.

Habit—Hemimorphic pyramidal [$50\bar{5}2$] and [$10\bar{1}1$]. Short prismatic to tabular [0001]. Usually striated horizontally on [$10\bar{1}0$] and [$10\bar{1}1$]. Often fibrous or columnar; as concentrically banded crusts.

This information attempts to describe the general appearance of all crystals of wurtzite ever seen.

From the above, the crystals are pictured as being normally somewhat flat, with the two ends of the crystal having different sets of faces. Certain faces are described as being grooved or striated. The appearance of masses of the mineral is given so that it can be recognized when not in single, measurable crystals.

Structure cell—Space group C6mc a_0 3.811, c_0 6.234, a_0 : c_0 = 1.6358 contains Zn_2S_2.

This is structure cell data acquired by X-ray. It is perhaps the most tersely stated of all and yet the most significant information obtainable about the species. It describes the internal structure, detailing the kind of lattice and its symmetry. It gives, as well, the shape and size of the unit cell and the atoms that are contained in each unit cell. The structural arrangement is hexagonal (C6mc)

FACING: *Stromboli, still an active volcano, erupts on 30 August 1842.* RIGHT: *At the Dinosaur National Monument in Utah, visitors may watch a working paleontological occurrence. Here, part of the bone bed is being exposed.*

which agrees with the crystals measured with the goniometer. The length of the vertical dimension of the unit cell is actually 6.234 angstroms ($c_0$6.234) and the horizontal dimension is 3.811 angstroms ($a_0$3.811). When these lengths are compared with each other, the vertical dimension comes out 1.6358 times as long as the horizontal. This compares very well with the 1.6349 obtained from crystal measurement given before. When the chemical composition and density measurements are combined with the dimensions of the unit cell, it is possible to calculate that there are two zinc atoms and two sulphur atoms in each unit cell (Zn_2S_2).

There are several kinds of X-ray instruments currently in use. The X-rays are usually produced by a glass-jacketed tube containing a filament and a water-cooled target.

Forty to fifty thousand volts are supplied to the tube, and electrons streaming from the filament are smashed into the target by this high voltage, causing it to produce X-rays. The kinds of X-rays produced depend on the kind of target in the tube.

Usually a copper target is used, but for iron, cobalt and manganese minerals these X-rays cause undesirable fluorescence. An iron target gives better results. The tube has two or four small circular windows, about half an inch in diameter, covered by very thin plates of beryllium metal or mica. These materials are quite transparent to X-rays, even though they may look opaque to the eye.

When the beam of X-rays is directed at a solid, much of it passes through the solid with no alteration, some of it is scattered, and some is converted to heat or other kinds of energy. It is the scattered X-rays that are of interest because they are the ones that have hit atoms on the way through. A picture of these scattered X-rays is taken by proper placement of a photographic film. A large number of atoms all uniformly placed in the structure will scatter the X-rays in the same direction and reinforce their image on the film. The others are scattered in various directions and produce no combined mark on the film. This means that every different plane of atoms in the structure leaves its print on the film. By careful

In thousands of years, Monte Rosa Glacier in Switzerland will have ground a U-shaped bed out of its valley. Like all glaciers, Monte Rosa moves through partial melting, and on its way picks up and deposits material in moraines.

DEBENT IGNARI RES FERRE ET POST OPERARI QVATVOR INSERTA NATVRIS IN NVBE REFERTA
IVS LAPIDIS CARI VILIS SED DENIQS RARI NVLLA MINERALIS RES EST VBI PRINCIPALIS
VNICA RES CERTA VILIS SED VBIQS REPERTA SED TALIS QVALIS REPERITVR VBIQS LOCALIS

measurement of the film markings and suitable mathematical treatment of the measurements, the entire internal structure is revealed.

The X-ray method most often used is the powder diffraction procedure. The source of the X-rays is as described above, but the camera to hold the sample and film is unique. The camera is a flat, hollow metal cylinder, one end of which is a removable lid. It is very carefully machined to exact dimensions, since its uniformity and the size of its diameter are crucial in the final film measurement. There is a hole on one curved side for the entrance of X-rays, and a hole opposite for the exit of most of them. Tapered metal tubes fit in these entrance and exit ports to guide the X-rays to and from the sample,

which is mounted on a rotating spindle at the exact centre of the camera. The film is a long, narrow strip that fits flat against the inside wall of the cylinder. It has two holes that fit over the entrance and exit port tubes.

In operation, the sample is mounted, the film is loaded in the darkroom, and the lid is replaced. As the sample spindle is turned by a motor-driven belt, the entrance port of the camera is placed against the X-ray source, and the exposure proceeds, taking several hours. When the film is developed, it has a series of matched curved lines running across it. These lines represent atomic planes; for each species their pattern is characteristic and will nearly always be different from that of any other species. The films

can be indexed, filed and used much like fingerprints.

More information, and more precise measurement with simpler mathematical treatment, can be obtained by using a precession-type camera, perfected by Dr. Martin Buerger of the Massachusetts Institute of Technology. It refines a method first used at the beginning of X-ray crystallography. Making measurements and calculations from these early films was difficult because there was a distortion of distances on the film. This distortion increased the farther the dots were from the centre. Buerger overcame this difficulty with his precession camera, a mechanical contrivance that moves the film through strange gyrations during the exposure. Although exposures may run for two or three days, the resulting film is free from bothersome distortions.

Another apparatus often used in crystallography is the X-ray diffractometer. This instrument, performing the same function as the powder camera method, is more expensive than some of the others and requires a larger mineral sample. However, it gives very precise recordings of the spacings between planes of atoms, which are measured with some difficulty on the powder and precession camera films. It can also give the precise intensity for X-ray reflections from each plane. The sample is ground and smeared out with an adhesive on a glass slide. The slide is then placed in line between the X-ray tube and a Geiger counter. The counter moves in an arc about the sample as the X-rays come through it, picking up the reinforced X-rays that are scattered by the sample at different angles. The Geiger tube is connected through electronic circuits to a recording device, which automatically

FACING: '*The Alchemist's Laboratory*', *after Breughel, sixteenth century. Alchemists dreamed of converting common metals into gold.* ABOVE: *Despite new electronic devices now available in the mineralogical laboratory, the skilled analytical chemist is indispensable.*

draws a chart, showing peaks on the graph when reinforced X-rays scattered from a plane of atoms hit the counter. Since the chart is on a larger scale and more easily read than a film, the results are in a more practical form.

There are other X-ray devices in use, as well as several modifications of the devices described here; the X-ray crystallographer has available to him a good deal of special-purpose equipment.

Chemistry—Zinc Sulphide, ZnS. Contains, in addition, some iron (8 per cent maximum) and cadmium (3.66 per cent maximum).

The description of the chemical composition of wurtzite shows that it is not a pure compound of zinc and sulphur but contains considerable iron and cadmium, as impurities or in substitution for some of the zinc atoms.

Chemical analysis of minerals was widely practised for decades before anything was known about their structures. Naturally, the methods of chemistry have progressed along with the rest of the science. Many analytical chores are now performed by electronic equipment. But the other extreme is the simplest kind of chemical investigation: the spot test. In this kind of test the chemist is checking for a specific element or group of elements he thinks might be present in the sample. Taking a small part of the sample, he adds a few drops of an indicated chemical solution and watches to see if the expected results are produced. For example, when a drop of vinegar or hydrochloric acid is placed on a piece of calcite, it responds by giving off quantities of bubbles of carbon dioxide gas, proving that it is one of the carbonate family of minerals. Very precise spot tests, much more reliable and useful than this one, have been devised for many of the elements and their compounds.

But nothing can replace a complete, thorough, painstaking analysis by the chemist in his laboratory, where he has a wide array of glassware, chemical materials, weighing instruments and other aids. As he goes carefully through his procedures, step by step, he discovers which elements are present in his sample, and in what quantities. His success depends on his knowledge of techniques, his judgment in deciding which techniques to use, and his skill in carrying them out.

Obviously his task would be very much simplified and considerable time and material would be saved if the chemist knew which elements were present before he started. He could then determine the quantities of each element and be secure in ignoring the possible presence of others. In all modern laboratories this information can be obtained by spectrographic methods, among others. The spectroscope is based on Isaac Newton's original discovery that a beam of light is split into various wavelengths as it passes through a glass prism. In actual practice a glass prism can be used in the instrument, but it is more usual to use a diffraction grating. The sample to be examined is powdered and placed between two carbon rods. An electric arc is struck across the carbons and the sample is vaporized in the arc. As the atoms in the elements of the sample are heated in the arc they glow brightly to incandescence. The wavelengths of light given off by each kind of atom are different. The prism or diffraction grating sorts them out. After sorting, they fall on a photographic film and are recorded. By comparing the developed film with a standard measuring film, all the wavelengths can be identified; the elements that caused them to appear are known, as are the quantities of each.

One of the most complex chemical instruments to be devised in recent years is the extremely expensive electron microprobe. It can perform

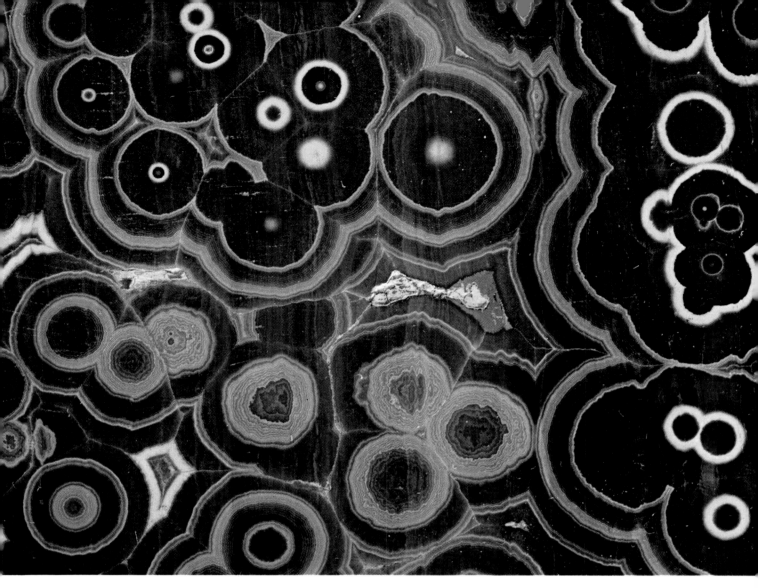

TOP: *Agate from Kambambi, Madagascar.*
LEFT: *Wavellite from Germany.*
ABOVE: *Iridescent fossil shell,*
South Dakota.

chemical analyses, impossible by wet methods, at a very high speed. The sample to be analyzed can be less than one micron in diameter (three hundred-thousands of an inch), too small to see without magnification. The sample is first prepared by grinding it flat, polishing its surface, and coating it with a very thin layer of graphite. Then the probe shoots a high-intensity, tightly focused beam of electrons through a vacuum at the sample, which is also in the vacuum chamber. The beam hits the target and excites the atoms into giving off their own particular kind of radiation. The radiation is collected and sorted by devices called curved crystal spectrometers. The sorted X-ray wavelengths are counted automatically and the information fed directly from the counters into computers. The information can be typed out by automatic typewriter as a set of numbers, or recorded in graph form by an automatic inking pen. If the beam is used to scan, like radar, across a larger sample, a picture will appear on a special oscilloscope tube, like a television tube, and can be photographed. If the scanning is set for a single element, iron for example, the photograph will show only those sample areas containing iron. No matter how the information is extracted, mathematical treatment of the results will give a good, serviceable chemical analysis in a short time.

Optical properties—Uniaxial positive. Indices O 2.356, E 2.378 for sodium light.

This information, obtained from the behaviour of wurtzite with light, tells us that wurtzite is either hexagonal or tetragonal and that it has two different indices of refraction, 2.356 and 2.378, as expected for a doubly refracting substance. The light used to test the sample was from a sodium lamp, commonly used for this purpose because it has restricted wavelengths of light rather than the whole range of white-light wavelengths. Since the amount of refraction differs with different wavelengths of light, using only certain restricted wavelengths gives precise refraction measurements rather than a whole range of measurements for white light.

The petrographic or polarizing microscope was developed in the nineteenth century. Even with the advent of X-ray equipment, the polarizing microscope remains one of the most important basic tools for the mineralogist. Its importance is based on the fact that the effect of solids on light is a direct result of their structures. A knowledge of their optics, then, is an important key to a complete knowledge of the structures of minerals.

The polarizing microscope differs from ordinary microscopes in several ways. It does much more than merely magnify the specimen. Beneath the stage of the microscope, upon which the sample mineral sits, is a calcite crystal cut and polished in such a way that light coming through it, unlike normal light, vibrates in a single direction—or is polarized. Often a sheet of polaroid is used in place of the calcite crystal. The polarized light passes through a hole in the stage and up through the sample. Samples to be examined may be tiny mineral grains or 'thin sections' of rocks. A thin section is prepared for this microscope from a small slab sawn off the rock and then ground so thin that light will come through it. An average thickness would be about thirty microns (not quite one-thousandth of an inch). Farther up the microscope tube is another calcite crystal just like the polarizer down below. The two crystal polarizers, called nicol prisms, are placed so that one passes light vibrating in one direction and the other passes light vibrating only at right angles to it. This means that without a sample in place the observer sees a black field through the tube. No light can pass. When isometric mineral samples or glass samples

FACING: *The author with the Smithsonian Institution's new electron microprobe, which returns in minutes analyses that could take weeks by older methods.* RIGHT: *A small mineral sample is microprobed for distribution of various elements.*

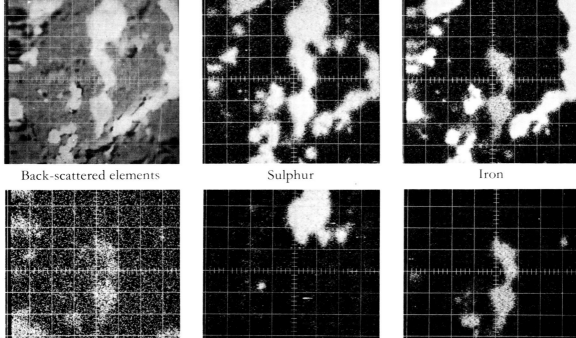

Back-scattered elements

Sulphur

Iron

Magnesium

Calcium

Manganese

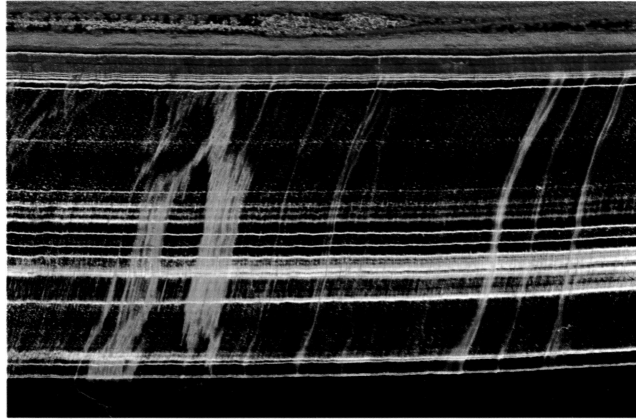

are placed on the stage between these crossed prisms, the field remains dark. Other mineral grains of other crystal systems, because of their light-refracting abilities, will be illuminted in the field.

The polarizing microscope is quite flexible. It can be used to determine the refractive index of a mineral, to reveal the presence of twinning in crystals, and to sort out minerals of different crystal systems. Grains that show no double refraction in any position are either glass or isometric minerals. Other grains that show no double refraction in only one position are tetragonal or hexagonal. Still others that show no double refraction in two different directions are orthorhombic, monoclinic or triclinic minerals.

Physical properties—Cleavage [$11\bar{2}0$] easy, [0001] difficult. H $3\frac{1}{2}$-4, G. 3.98, 4.1 [artif.], 4.1 [calc.]. Lustre resinous. Colour brownish black. Streak brown.

This gives additional general information about physical properties. From it we learn that wurtzite has a good cleavage in one direction and a weak cleavage in another. It tells us that the hardness is between $3\frac{1}{2}$ and 4—a fairly soft mineral. The specific gravity was measured on natural material to be 3.98, which compares quite well with measurements made on synthetic material at 4.1. In turn, the specific gravity of synthetic material is exactly what it should be, as predicted from X-ray information. Lustre, colour and streak complete the description.

Any mineral having all the crystallographic, X-ray, chemical and physical characteristics described here must be wurtzite and no other. The use of the instruments and techniques described, as well as many others, helps to insure the accuracy of the information.

As with any large group of professional people,

there are a number of organizations catering to the interests and needs of geologists. In the United States the Geological Society of America is an all-inclusive grouping of professional geologists. Other, more specialized, societies include the Geochemical Society, Paleontological Society, Society of Vertebrate Paleontologists, and the one that involves mineralogists per se, the Mineralogical Society of America.

The Mineralogical Society of America is composed of members working with materials that exist as solids. Those working with natural products are in the majority, but there are also many members who work with synthetic substances because the theories, laboratory techniques and methods of

FACING, TOP: *Millerite, a nickel sulphide in iron ore, from a defunct iron mine in Antwerp, New York.* LEFT: *From Griqualand, South Africa, a pseudomorph of quartz after crocidolite.* ABOVE: *An exceptional specimen of variscite.*

study overlap. From a small beginning at Harvard in 1912, the Mineralogical Society of America membership has grown to about twenty-five hundred. This seems a small number when compared to the many thousands of American amateurs in clubs peripherally involved with mineralogy. Nevertheless, the Society is the strongest, most active professional mineralogical group in the world.

At its annual meeting, the Society presents the Roebling Medal for outstanding achievement in mineralogy and petrology, one of the greatest honours a mineralogist can receive.

In 1920 the Mineralogical Society of America took over sponsorship of a small magazine previously published by other groups, and the *American Mineralogist* began to prosper. In 1926 it was endowed handsomely by Washington A. Roebling,

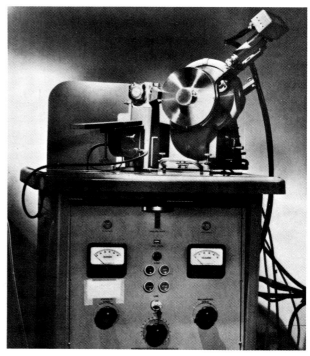

an amateur interested in mineralogical affairs. Later grants of money came from the Geological Society of America. Today the publication is internationally significant, with over two-fifths of the published copies going to subscribers outside the United States. It averages more than fifteen hundred pages a year and its contents are of high scientific quality.

The Mineralogical Society of Great Britain, founded in 1876, publishes its own journal, the *Mineralogical Magazine*. In collaboration with the Mineralogical Society of America it publishes *Mineralogical Abstracts*. This is a periodical that gives very short reviews, or abstracts, of all scientific papers of concern to mineralogists that are being published in all the world's journals. The abstracts keep the mineralogist informed of recent developments.

FACING, TOP: *X-ray powder-technique camera films atom patterns of powdered, rolled samples.* LEFT: *Similar but more precise data is given by X-ray diffractometer.* ABOVE: *A sample is placed in a spectroscope for arcing.*

Minerals at Work

8

Space rockets, capsules and the hardware to launch them, supersonic jet airliners, nuclear power generators, and a host of other complicated fabrications have in recent years had a drastic impact on our exploitation of the mineral kingdom. The demand for rare metals, some previously known only as laboratory curiosities, has accelerated. New kinds of metal alloys, new kinds of crystals, and other substances with peculiar and unique properties are constantly being sought to satisfy the needs of an imaginative and rapidly expanding industrial technology. Most of these materials are extracted directly from the mineral deposits of the world or are manufactured from mineral products.

Everything man now uses except food, other agricultural products, and clothing, has been mined from the earth. Considering that all this mining underpins our industrial culture, it is amazing how little general interest or knowledge there is about it. Such matters are considered too technical and the proper business only of scientists, engineers and technicians. Except in time of crisis, most governments show little interest in mineral exploration and study. Consequently, the exploitation of the world's mineral resources is left to drift, buffeted by political whim and the fluctuations of the market. Nevertheless, our industrial civilization and its advancement depend completely on mining and the practical use of the products of the mineral kingdom. The story of the search for minerals, their extraction from the earth and their application in industry is a fascinating one.

IRON: Homer mentions iron throughout the *Iliad*, his epic of the Trojan wars. For example, he tells of Achilles offering an iron quoit as a prize at the funeral of Patroclus. The ancient Greeks knew the metal, then, and considered it rare and precious. Iron objects have been recovered from Egyptian tombs dating from 3500 B.C. and earlier. By the time of the Hittites in Asia Minor, it was being worked in sufficient quantity for practical use. In 600 B.C., according to Byron's 'Destruction of Sennacherib':

The Assyrian came down like a wolf on the fold,
And his cohorts were gleaming in purple and gold;
And the sheen of their spears was like stars on
the sea . . .

It is certain that those brightly shining spears were made of iron; these warriors were completely outfitted with iron armour and weapons.

It is safe to say, then, that mankind has had a practical technology for ironworking for at least three thousand years. During that period the use of iron, replacing stone and bronze, caused such drastic changes in man's life and activities that it has become known as the Iron Age. We are, perhaps, still in this age.

There is enough iron in the crust of the earth to account for most of the yellow, red, brown and green colours seen in exposed soil and rocks. The amount of aluminium in the earth exceeds that of iron, but aluminium is more difficult to extract from its ores. Iron, therefore, is the commonest commercial metal.

The world now has a production of iron and steel of various compositions and purities— produced by several different processes—exceeding four hundred million tons a year. In such quantity the costs of production are minimized; certain kinds of raw iron products now cost as little as a penny a pound.

To produce these prodigious quantities of iron and steel requires literally mountains of ore. The problem would be much simplified if iron were the only raw material needed. But to charge one of

PRECEDING PAGES: *Minerals are used in new ways to meet the intricate new demands of space technology. Light-sensitive plates of silicon, which convert sunlight to electricity, coat a satellite in flashing blue.*

the enormous fiery blast furnaces that melt the iron from its ores usually requires three other important raw materials. For every four tons of iron ore smelted in the furnace, it is necessary to add about one ton of limestone to remove the silicate mineral impurities, two tons of coke to supply the heat and to keep the right amount of carbon in the mixture, and nine tons of air to keep the chemical reaction going. The air is free, but the coke must be made from coal and the limestone must be mined. The iron ores themselves must be mined. The United States and Russia are the major iron ore producers, followed by France. Supplies of higher-grade ore in the U.S. are diminishing and more iron ore is being imported from Canada and South America.

There are three major ores of iron: magnetite, hematite and limonite. Hematite is most commonly used because of its more general availability, purity and high iron content. Hematite, the red ore of iron, is an iron oxide (Fe_2O_3) and theoretically contains as much as 70 per cent iron. As it is usually found in its deposits mixed with impurities, the percentage actually is lower, and deposits containing as little as 25 per cent can be worked with profit. The most common impurities are quartz, clay, limestone and traces of the elements manganese, titanium, sulphur, arsenic and phosphorus. The presence of titanium and manganese can enhance the value of the ore, but sulphur, arsenic and phosphorus are undesirable in the finished product.

In the district around Lake Superior lie the richest and most productive hematite mines in the world. The largest of these are the iron ranges in Minnesota, but there are others of the same type in nearby Michigan. Unfortunately, the best of these deposits were stripped off during World War II, and current

From the time of the Hittites, iron was used for armour and the tools of war. In this Renaissance painting, Giulio Romano depicted Achilles receiving from Thetis the metal arms forged by Vulcan.

鐵潮泛灰　鐵方流　生成流此管　墮子銅　板生鐵

點錫　勻錫　流入鐵盤

鑞弓　空管　入此水潮　青田

operations are plagued with lower-grade ores containing excessive quartz. Processes have now been developed, however, to make mining and smelting of these high-silica taconites profitable.

There is considerable professional debate over the sources of these deposits. Some geologists consider them simple sedimentary beds. Others believe evidence proves that the iron and silica were carried upwards by magmatic water and deposited on the floor of an ancient sea. Much of the silica then was leached, or dissolved away, converting low-grade deposits to usable ore.

Major Russian hematite deposits are located near Krivoi Rog in the Ukraine. Other new and important deposits have been found elsewhere, such as one on the Quebec-Labrador border in Canada, containing an estimated six hundred million tons of high-grade ore.

Black iron ore—magnetic ore—is less plentiful. It consists of the mineral species magnetite, which is also an oxide of iron (Fe_3O_4) and theoretically may contain up to 72 per cent iron. There are workable deposits of magnetite in the United States and in Russia, but the largest and best known are at Kiruna and Gällivare in northern Sweden. Swedish steel, made from ore mined at these deposits, has long had an excellent reputation. The mines are very large, open quarries and the deposits contain reserves estimated at a thousand million tons.

The common, yellow-brown bog-iron ore, or limonite, is of sedimentary origin and lower grade, with a maximum theoretical content of 60 per cent iron. Limonite is a mixture of iron oxides with varying amounts of water in their compositions. It is often very impure because it is mixed with other sedimentary materials such as clay, sand and gravel. The best-known and most productive limonite mines are located in the Alsace-Lorraine

district of France and produce ore suitable for smelting.

All iron, which is smelted in great blast furnaces, is run off as a molten liquid to be cast in bars or pigs. From these it will eventually be converted into cast iron, wrought iron or some kind of steel.

To make cast iron, the pigs are remelted and poured into moulds to produce the proper shapes. Because cast iron resists heat, it can be used for boilers and fireplace grates. Since it does not rust easily, it is widely used for water pipes and gutters. It also is frequently used for large, basic parts of heavy machinery.

Wrought iron is low in carbon and is heated and worked or drawn while hot. This process makes it very tough and resistant to shock, unlike cast iron, which is brittle. Because of its toughness,

FACING: *Metallurgy in ancient China.* TOP: *Iron being smelted. The ore was puddled (at left) to purify it.* FAR LEFT: *Smelting tin.* LEFT: *Mercury smelted out of cinnabar.* ABOVE: *A Bessemer furnace in 1856; the process is basically the same today.*

195

wrought iron can be worked into chains, bolts, nails, pipes and hinges, which are subject to abuse.

There are two methods for converting pig iron to steel. Of these, the Bessemer process accounts for a small proportion of the steel made. The rest is made by the open hearth method. Between fifteen and twenty tons of liquid pig iron can be handled at one time in the egg-shaped Bessemer converter, while the open hearth method can handle between fifty and two hundred tons in each batch and is easier to control because the process takes much longer.

Smaller batches of special steels are made in electric furnaces, where they are heated for the addition of certain desirable elements such as nickel, cobalt, molybdenum and vanadium. These give steels greater elasticity, greater hardness, high corrosion resistance, magnetism and other desirable qualities.

ALUMINIUM: Six times more iron in its various forms is produced each year than its nearest metallic competitor, aluminium. The competition has been keen, and aluminium has reached its prominent position in the metals market in less than eighty years. There are many uses for iron in which

aluminium cannot be substituted, but many new applications have been developed for this common metal.

Since aluminium is much more plentiful in nature than iron, it is surprising to discover that it was unknown to the ancients. It was not until 1825 that the Danish chemist Hans Christian Oersted found a way to extract aluminium from its ores. Even then it was not possible to prepare it in quantity until 1850. At that time, although still difficult to prepare, it could be bought in small quantities at about the same price as gold. By today, the price has been reduced so much and technology so improved that in many applications, such as extrusions, aluminium is cheaper than any other material.

In the year 1886 Charles Martin Hall found a way to separate aluminium away from its ores. The best ores of aluminium are those in which it is combined with oxygen with very strong chemical affinity. Bauxite, the best and commonest ore, is such an oxide with a small amount of water in its composition.

Hall's method made it possible to separate the tightly bound aluminium on a large scale and at a low enough cost to make the metal economically competitive. He discovered that aluminium oxide (Al_2O_3) would dissolve in melted cryolite (Na_3AlF_6). Cryolite is a mineral species that has been mined in commercial quantities at Ivigtut in western Greenland. It melts at rather low temperatures, and is excellent for the purpose.

In Hall's process, the aluminium oxide is dissolved in melted cryolite in a carbon-lined container, and carbon bars are inserted into it. An electric current is run through the solution between the carbon bars and the carbon lining. The electric current drags the aluminium atoms away from the oxygen

ABOVE: *An artist's impression of coining in the fifteenth century. The word coin derives from the die—cuneus—that stamped the coins.* FACING: *Sixteenth-century decorative pewter plate; the tin-lead alloy was commonly used for utilitarian objects.*

atoms. It settles as molten metal on the bottom of the container, where it can be tapped off as it accumulates. By the steady addition of aluminium oxide, the process can be kept going continuously until the carbon pieces need replacement. Of course, the process consumes fantastic amounts of electricity. Aluminium manufacture is therefore one of the major consumers of electricity in industry today.

The trouble with the early aluminium produced was that the metal lacked strength, was soft and extremely brittle, and had a tendency to flake and corrode. Metallurgists have found ways to increase its purity and to alloy or mix it with other metals for strength. As a result, aluminium can now be made as strong as steel, highly resistant to corrosion, and it is even available in several colours. Duralumin is a good example of its alloys. It consists of 4 per cent copper, 1 per cent manganese, 1 per cent magnesium, and the rest aluminium, and although not quite as strong as steel, is a satisfactory substitute, particularly because it has only one-third the weight of steel.

Because aluminium is lightweight and an excellent

conductor of electricity, it is used for long-distance electrical transmission wires. Ordinary copper wires are quite heavy. Aluminium makes it possible to have larger wires that carry more current, without causing a weight-supporting problem.

Like iron, aluminium does corrode. However, there is a major difference. As the oxygen of the air attacks and combines with the exposed aluminium surface, it is held tightly by the strength of the attraction between them. This quickly coats the surface with a very thin, very tight-packed, transparent layer of aluminium oxide. Since no more oxygen can get through this protective layer to attack the fresh aluminium underneath it, corrosion stops. This layer of aluminium oxide also protects aluminium from the chemical attack of foods. Since it also conducts heat well it is ideal for cooking utensils.

GOLD: Among the Egyptians and other cultures in the cradle of western civilization gold was used at least as long ago as 4000 B.C. It is used today as the basis of the world's money system, for ornamentation and for jewellery. For centuries, then, it has had a high intrinsic value.

Gold is very well qualified to be a coinage metal. It is highly resistant to corrosion and chemical destruction of almost any kind. Also, it is soft and easily worked. Scarcity helps to keep its value high, but enough can be found to satisfy monetary and industrial needs. Silver, the only other true coinage metal, is more common and does not have the same degree of durability as gold; it takes a second position. All coins made of other metals are merely money tokens based on the value of stored gold and silver.

Gold is a heavy substance. A piece of it is three times as heavy as a piece of steel the same size. A small block of solid gold the size of a normal pack of cigarettes would weigh almost four and one-third pounds.

Gold is chemically inactive. Evidence of this is its occurrence in nature as a free metal, uncombined with other elements. Even in the laboratory, gold is not attacked by the common acids. The ancients found that a mixture of hydrochloric acid and nitric acid would dissolve it, though with some difficulty. They named this mixture *aqua regia*—royal water—since it could dissolve gold, king of metals. Although gold is so chemically inert that it cannot possibly corrode, it is a very soft metal and is usually alloyed with copper, silver or nickel to make it hard enough for normal uses. Besides being soft, it is malleable, which means it can be easily beaten into sheets or drawn into wire. One ounce of gold can be beaten thin enough to make sheets that will cover an area thirteen feet square. One-third of a million sheets of this gold leaf would make a pile only about an inch high. Since gold can also be drawn into wires as thin as one two-thousandth of an inch in diameter, and has very high electrical conductivity, it finds critical uses in the electronics industry.

Sometimes the alloys of gold are more useful than the pure metal. This is especially so in coinage, ornaments and other objects that suffer rough use. When absolutely pure, gold is labelled—for historical rather than scientific reasons—24-carat gold. Most jewellery gold is alloyed with copper to give it strength and hardness. Good quality jewellery is of fourteen parts of gold to ten of copper (14 carat) or eighteen of gold to six of copper (18 carat). Inexpensive jewellery that turns green or black after exposure to the weather or human skin contains too much copper and too little gold. Notice that the unit called the carat used to express the purity of gold is an entirely different unit from the carat used as a measure of the weight of gemstones (see page 73). Coinage gold is usually about 22 carat, the other constituent being copper. Dental gold varies a bit but contains from 65 to 90 per cent gold and 5 to 12 per cent silver. Gold is also graded as to its 'fineness', based on how many parts out of one thousand are gold. An alloy of 85 per cent gold and 15 per cent copper would be 850 fine.

FACING, FAR LEFT: *Thousands of formulas exist for making glass; quartz sand is basic to all.* LEFT: *Charles M. Hall invented the electrolytic production of aluminium.* RIGHT: *Extracted aluminium is poured into ceramic moulds for casting.*

To extract gold from its crushed and roasted ores, the mercury amalgam method described in Chapter 4 is commonly used. Surprisingly, however, a quarter of the gold mined in the United States comes from placer mining. The gold-bearing gravels so mined have come from the weathering and breaking-up into gravel deposits of the original gold-bearing mother lodes or veins. Because gold is so heavy, it tends to settle out and concentrate itself at the bottom of these deposits, so that a small tonnage of gravel concentrates can yield a high percentage of gold. Sometimes the gold nuggets in placer deposits are startlingly large. Historical reference is always being made to the 'Welcome Stranger' nugget found at Ballarat, Australia, in 186. It weighed a glorious 2280 ounces.

South Africa is currently the world's foremost producer of gold. Some of the richest mines occur in the Rand district around Johannesburg. Rich as they are, the Rand ores contain grains of gold too small to be seen by the naked eye. To the delight of the owners and operators, they also produce quantities of uranium.

The most favourable place to find gold, other than in placer deposits, is within a mile or so of a granite intrusion. The gold will have been deposited there by hydrothermal action. Some of the remarkable deposits at places such as Cripple Greek, Colorado, and Goldfield, Nevada, resulted from low-temperature hydrothermal action near the surface. This kind of vein is spectacular and fantastically rich, but such deposits are short-lived and do not persist at depth.

QUARTZ: This very common mineral has been put to many uses—some of them rather surprising. Quartz is found in nature as grains or masses that are white, grey, or sometimes highly coloured. It also occurs in clear glassy crystals and masses that can be of considerable size. The largest single crystal on record was found near Itapora in Goiás, Brazil. It was twenty feet long, and its estimated weight was over forty-four tons.

Much of the sand of ocean beaches is composed of quartz grains. Quartz in the form of sand has valuable direct use in the building trades when it is mixed in mortar and cement. In powder form, it is mixed in paints to give them heavier body; it is used in scouring soaps and on sandpaper to produce scratching or abrasion. It will eventually wear away most substances, though this does not apply to extremely hard metals or minerals.

The glass industry also consumes large quantities of quartz in the form of sand. There are perhaps as many as fifty thousand different formulas for making glass, each one giving a product with slightly different characteristics. Typical of these is a simple soda-lime glass made by fusing together pure quartz sand with sodium carbonate and calcium carbonate. Bottle glass is a simple glass of this type, with a greenish colour arising from traces of iron impurities. There is also a quartz glass made from pure sand alone—very expensive because it is difficult to prepare. A mixed-composition glass will melt, and can be worked, somewhere between 600 to 900 degrees Fahrenheit. Fused quartz glass, however, melts at 1500 degrees, and must be kept at that temperature while it is being worked. However, it has many advantages over other kinds of glass. It is transparent to ultraviolet light, which is stopped by ordinary glass. Sun lamps, and other lamps used as sources of ultraviolet light, must have bulbs made of quartz glass.

Fused quartz glass also withstands rapid changes in temperature without breaking. This characteristic is called a 'low coefficient of expansion', meaning that there is very little expansion or contraction

*Most steel is produced by the open-hearth process,
easier to control than the Bessemer. In a steel mill at
Hamilton, Ontario, slag, which contains the dissolved
impurities, is poured from a typical open-hearth furnace.*

with changing temperatures. Even when the glass is unevenly heated or cooled, so that one part is hotter than another, there are practically no expansion stresses. Consequently, quartz glass is used extensively in the laboratory in the form of tubes, rods and containers.

In one unusual process, melted quartz is forced out through a stream of high-pressure hot air. This spins the quartz out into a fluffy mass of fine fibres, from which a quartz paper is manufactured. This fibre, mixed with clay, is run through a paper-making machine to become quartz paper—a good electrical insulator because it will not conduct a current, nor will it break down when placed in a strong electrical field. At the same time it can withstand temperatures as high as 3000 degrees Fahrenheit.

Quartz also has some interesting electrical characteristics. It is a piezoelectric mineral, meaning that crystals of quartz, when squeezed, develop electrical charges on certain surfaces. The process also works in reverse. If an electrical charge is applied to a crystal, internal stresses develop and it changes shape.

When free of bubbles, cracks or flaws, and free of twinning in its internal structure, quartz is often used for piezoelectric devices. A disc or rectangular plate, cut from the quartz in very exact directions with carefully controlled dimensions, is mounted between two electrical contacts and placed in a protective tube. This so-called oscillator is then introduced into an electronic device such as a telephone circuit or a radio apparatus—circuits that have an electric current pulsing through them alternately back and forth thousands or even millions of times per second. As this alternating current hits the quartz plate, the plate expands and contracts rapidly. Its dimensions have been chosen so that it vibrates—oscillates—at the same rate as the circuit current. The current and the quartz crystal plate must operate in unison so that, with the plate unable to change its dimensions, it keeps the circuit from alternating at anything other than a fixed rate. The device then becomes known as a frequency-control oscillator. It is responsible for the fact that every time you tune in your favourite radio station it is at the same place on the dial.

With the tremendous advances made in electronic equipment during World War II, quartz oscillator plates were critically needed. Most of our supply came from Brazil. A good part of this electronic quartz is now grown synthetically, but large amounts are still imported.

URANIUM: In 1789, uranium was discovered by the chemist Martin Klaproth in black pitchblende from the Czechoslovakian mines at Jachymov. He

LEFT: *The exhausted west pit of an Australian uranium mine, Kathleen Uranium Limited.* FACING: *One of the end products: the mushroom-shaped trademark of the twentieth century—the cloud of an atomic explosion.*

named the element after the planet Uranus, which had recently been discovered. Originally the Jachymov mines had been operated for their silver ores, which eventually ran low. They were then converted to lead mining, but little thought was given to the pitchblende they contained.

It was in this same black material that Becquerel discovered radioactivity in 1896, and Marja and Pierre Curie discovered polonium and radium in 1898. Even though by then acutely interesting to science, uranium and its ores were still commercially unimportant.

In 1939 the entire picture changed with the discovery of atomic fission. O. F. Hahn of Germany, repeating some of the atomic research of Enrico Fermi of Italy, reported certain unexplained results. These were correctly interpreted in 1939 by Lise Meitner and O. R. Frisch, who had fled from Nazi Germany and were working at the laboratory of Niels Bohr in Denmark. They were farsighted enough to recognize in atomic fission a phenomenon that would instantaneously release very large amounts of energy. Atomic fission is the splitting of the cores of certain kinds of atoms by bombarding them with other kinds of atomic particles called neutrons. These cores, or nuclei, form atoms of lighter elements as they split, and in the process release tremendous amounts of raw energy.

Uranium atoms are not all the same. They come in different weights. The scientist labels them as uranium 235, uranium 238 and uranium 239. The one that undergoes fission is uranium 235. Unfortunately, only one part in one hundred and forty of natural uranium is this kind.

Because all uranium atoms behave the same way chemically, there is no way to separate them by simple chemical sorting. Enormous factories and whole cities like Los Alamos had to be created to handle the complex processes for sorting out the uranium 235. Along the way a method was found for creating a new element, plutonium, from the uranium raw materials. Plutonium has the same fission characteristics as uranium 235, but to a greater degree. Furthermore, being a different element, plutonium has different chemical characteristics and can be chemically separated from uranium.

Surmounting the difficulties of handling and processing tonnages of these rare and highly radioactive metals was an unbelievable task for science and industry. But it was done, with results that in 1945 wiped out Hiroshima and riveted the attention of the world.

Before 1923, the deposits of carnotite, the yellow uranium vanadate mineral found on the Colorado Plateau in the United States, were the most important in the world. By 1929, however, rich deposits of mixed uranium minerals in the Congo had forced American producers from the market. Very soon afterward the world supply was heavily increased by deposits from Great Bear Lake, Canada. With the rapid change brought about by the uranium fission discovery, however, there were more than one thousand operating mines in the United States and Canada by 1961. In the next few years, uranium was over-produced, and production was cut back sharply at great financial loss. Currently, with the rapid development of nuclear power plants, the demand for raw materials has again increased.

RUBY: Much of the best natural ruby—a chromium-bearing variety of the mineral species corundum—is converted to gemstones. Off-coloured or flawed ruby is sometimes used for bearings in watches and other delicate instruments. It is excellent for bearings because corundum, being so very hard, will not be worn away by the constant motion of a gear or wheel and needs no lubrication, as would a metal gear turning in a metal socket.

Because it is so easy to manufacture ruby these days (see Chapter 3), most of the highly technical uses for it are based on synthetic material. However, the raw material for making ruby still must be mined and purified. The ore of this aluminium oxide (Al_2O_3), or alumina, is bauxite, and is the same ore from which aluminium metal is extracted. Man-made ruby is prepared for bearing manufacturers in long, thin, cylindrical rods of the required diameter for the finished bearing. These rods are sliced into little circular tablets of the desired thickness.

The impurities that cause red colouring in corundum have no importance in bearings, but they have a most significant effect on the electronic characteristics of the mineral. They are responsible for the most fantastic new use to which any mineral material is being put—the application of synthetic ruby in masers and lasers. The two words are rapidly becoming part of everyday vocabulary, even though what they describe is largely a mystery to the public. Maser is an abbreviation for 'microwave amplification through stimulated emission of radiation'. The device, often built around a synthetic ruby rod, can receive a weak radio signal and be stimulated by it into giving off its own radio signal of the same wavelength—but one thousand times more powerful. Such devices are of enormous importance in receiving weak television and radio signals bounced off communications satellites, like Telstar, and amplifying them for re-broadcasting. They also receive and amplify the very faint radio signals that radio telescopes receive from distant

FACING, TOP: *Orpiment, an arsenic sulphide, in rare crystals from Takab, Afschar, Iran.* RIGHT: *Cinnabar, a mercury sulphide, in unusually large, dark red, twinned crystals from Hunan Province, China.*

stars and galaxies, thus helping us to extend our knowledge of the heavens.

An optical maser is very similar to a radio- and television-type maser, except that it deals with light. It is called a laser—an abbreviation of 'light amplification through stimulated emission of radiation'. Lasers have been made using a variety of materials, including glass. A simple ruby laser has a cylindrical, laboratory-grown ruby crystal with one

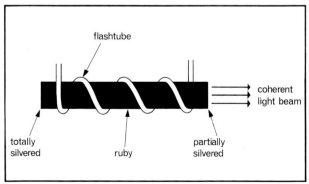

field that increases its power one thousand times more!

In addition to communication, the concentrated light power of lasers will probably one day help us to destroy missiles, bore holes in steel, make three-dimensional photographs, navigate satellites and fight cancer. It is already being explored as a new surgical technique.

SULPHUR: The most important single chemical substance used in a modern industrial country is sulphuric acid. In one way or another, it is involved in nearly every industrial process. Sulphuric acid is made from the element sulphur, together with air and water. In the United States, tremendous sulphur deposits are found in the rocks lying on top of great intrusive plugs of rock salt called salt domes. Such a dome may have in its capping tens of millions of tons of sulphur.

end silvered so as to act as a mirror, the other end partially silvered. Around the ruby is placed a coil-shaped flash tube. When this flashes, it excites the chromium atoms in the ruby into giving off a continuous, high-intensity, coherent light beam. (A coherent beam is light of one wavelength vibrating in phase, unlike white light with its mixture of wavelengths and phases.) The silvered ends reflect the light back and forth within the crystal, amplifying it, and a fraction of the beam emerges through the partially silvered end as coherent light.

This beam of light, as it shoots out the end of the laser rod, is so narrow and so intense that it can be transmitted over enormous distances without being lost. A beam of red laser light vibrates nearly a billion times as fast as ordinary radio waves. Radio waves are modulated, or modified, by radio transmitters to carry messages. It is possible also to modulate a laser beam for carrying messages. Because of its great vibration rate the laser beam could carry far greater quantities of information than do radio waves. And there is now a new technique for modifying the laser beam, using a powerful magnetic

These salt domes, located in Texas and Louisiana, pose a difficult mining problem. The deposits lie at considerable depth under quicksand and other rock layers. In 1894 a Philadelphia-trained chemist named Herman Frasch devised an ingenious method of recovering this deeply buried material. Three concentric pipes are forced from the surface down into the area of the deposit. The outer pipe carries down superheated water at about 360 degrees Fahrenheit. Compressed air, at 400 pounds per square inch, is carried down the innermost pipe. When the hot water reaches the deposit, it melts the sulphur, which is then forced up through the middle pipe by the compressed air. Since this middle pipe is surrounded by the hot water pipe, the sulphur remains hot and molten all the way up. At the surface, the sulphur, water and compressed air mixture froths out into towering rectangular vats, where it is allowed to settle, cool and harden. The sides of the vats are later removed, and the

FACING, TOP: '*Extraction of sulphur from pyrites by distillation*', Georg Agricola (1494-1555).
BELOW: *Cross-section of a simple ruby laser.*
ABOVE: *Yttrium garnet made at the Bell Telephone Laboratories.*

207

rectangular mountains of sulphur are mined with bulldozers, power shovels and other heavy equipment. About 60 per cent of the United States' supply of sulphur is obtained this way.

The other major source of sulphur is the iron and sulphur mineral pyrite, which is mined like most other solid ores. When the pyrite is roasted, some of its sulphur combines with air to form sulphur dioxide (SO_2), which is then used to make sulphuric acid. Enormous deposits of pyrite are found in the Rio Tinto area of Spain extending from the province of Seville westward to Portugal. The ores in this belt very likely represent a good half of the world's sulphur reserves.

GARNET: With sharper edges and greater hardness than quartz, garnet is especially valuable in the woodworking industry as garnet paper or cloth, replacing sandpaper. Natural garnet is also used for bearings in watches and other delicate instruments because it resists wear.

Of the several kinds of garnet, almandine has the largest industrial use. The only commercially important deposits of this iron garnet occur in the Adirondack Mountains of New York State. There the rock may be as much as 70 per cent garnet. The garnet is normally heat-treated to improve its hardness, then crushed, ground, separated by sieves, and graded according to grain size.

FLUORITE: Most highly industrialized countries have adequate supplies of fluorite—known in mining and industry as fluorspar. This is fortunate, since it is now recognized as better than limestone as a flux for making steel. About three-quarters of the world's annual production goes into open-hearth steel furnaces. There it melts easily and combines readily with the impurities in the steel. Once combined, it forms a very liquid and free-flowing slag that can be tapped off. For every ton of steel produced, about

Minerals have both frivolous and essential uses. FACING: *Colour in fireworks comes from minerals, among them antimony and strontium.* TOP: *Fluorite—most of it goes into steel production.* ABOVE: *Stibnite is the chief source of antimony.*

209

six pounds of fluorite is consumed. Fluorite is calcium fluoride (CaF_2); early metal workers recognized its ability to flow over impure molten metals and named it for the Latin word for 'to flow'.

Fluorite is used in the ceramics industry to make white enamel coatings for household fixtures and utensils. It is also a starting material for making a refrigerating substance used in automatic refrigerators, an easily liquified gas named sulphur hexafluoride. This gas weighs five times as much as air, is chemically inactive, odourless, colourless, non-poisonous, and will not burn. It is also used as an aerosol gas for spray insecticides, supplying the pressure to force the contents out of the container as it expands from a liquid to a gaseous state.

Pure, colourless fluorite is only single-refracting, transmits ultraviolet light easily, and has other optical characteristics that are ideal for certain kinds of special lenses. Lead fluoride (PbF_2) has been found to transform electrical energy into light. Fluorite is also the source of fluorine gas, used to prepare many fluorine compounds.

Fluorite is still recovered from old lead mine tips in the north of England. In the United States much fluorite is mined in the rich deposits of southern Illinois and neighbouring Kentucky.

ZIRCONIUM: In 1824, Baron Berzelius, a Swedish chemist, prepared the first small, impure sample of this silvery-white metal, which is difficult to extract because it combines easily with oxygen and nitrogen, carbon and silicon, and holds tightly once combined. Through improved purification methods, it has become a valuable industrial metal. It makes certain steels resistant to shock and fatigue, ideal for machinery that must operate under stress for long periods of time. Because its alloys resist common acids and chemical solutions, it is used, along with alloys of titanium, for bone-repair surgery.

One of the most interesting of industrial uses for zirconium metals is in atomic power installations, where its unique characteristics make it more useful than steel. One of these characteristics is thermal stability—ability to withstand the very high temperatures in nuclear reactors without breaking down. Because of zirconium's corrosion resistance, there is no rusting problem. Best of all, zirconium has what is technically called 'low neutron capture' characteristics. In order for a nuclear reactor to operate properly, there must be a carefully controlled internal ebb and flow of those fundamental particles called neutrons. Some metals have an

unwanted ability to capture some of these neutrons and to interfere with the operation. Zirconium has only one-thirteenth the capturing ability of iron and only one twenty-fifth as much as nickel metal.

The two most important ores of zirconium are the minerals zircon (zirconium silicate) and baddeleyite (zirconium oxide), recovered from beach sands or gravel deposits. The most productive deposits of zircon sand are found at Byron Bay, New South Wales. The best baddeleyite deposits are in the gravels and sands of Minas Gerais, Brazil. Other commercial ore deposits are found in Florida, India and western Africa.

FACING, TOP: *Vital in other industries, iron—much of it from hematite like this specimen—does not meet the requirements of space technology.* LEFT: *Beryllium is welded to aluminium and magnesium for NASA's Apollo.* ABOVE: *The satellite Pageos, to orbit 2600 miles above earth.*

Most zirconium ore is not used for extracting the metal but is converted to zirconia, which is a pure-white powdered zirconium oxide (ZrO_2). This white substance has a great many uses on its own, and it may also be readily converted to other important zirconium compounds. Ceramic materials made from zirconia can stand temperatures up to 5000 degrees Fahrenheit without melting. They can be made into ceramic parts that will stand much abuse. Zirconia ceramics have a low coefficient of expansion; they can be plunged at white heat into cold water without being damaged. They also have a high dielectric strength, meaning that they can act as

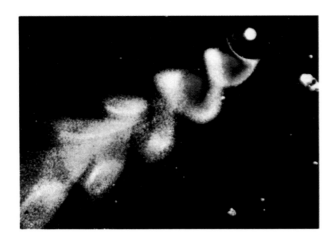

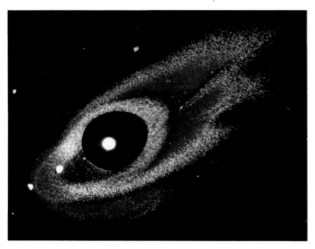

quality enamels. Zirconium carbide (ZrC_2) can be pressed into machine parts that are heat-resistant and more wear-resistant than metal. Zirconium sulphate ($ZrSO_4$) is used for tanning white leather. Sodium zirconium sulphate is used to make good-quality paper very white and very opaque. And zirconium carbonate is a salve for skin irritations.

STRONTIUM: As with so many metals pressed into modern service, strontium was unknown to the ancients. Not until 1790 was one of its minerals discovered at Strontian, Scotland. It is possible that its two major ores, the minerals celestine, strontium sulphate ($SrSO_4$), and strontianite, strontium carbonate ($SrCO_3$), were used before this date to produce fireworks with bright red colours. This is still one of the chief uses for strontium compounds.

But there are many others. Strontium titanate is made synthetically in clear, colourless crystals that are cut and sold as satisfactory gemstones. Another strontium compound is useful in the sugar-beet industry, combining readily with the sugar mixed with sugar-beet molasses to form a solid, strontium saccharate, which is then separated from the liquid molasses. By treating this substance with carbon dioxide gas, the sugar is recovered. The paint industry makes use of powdered celestine which, when added to paint, gives it brightness. There is also a peculiar compound called strontium fluoride (SrF_2) to which is added a small amount of the element samarium as an impurity. This compound is a valuable material for the lasers discussed earlier in this chapter.

SELENIUM: Selenium is a highly poisonous metal. Plants or foods grown in soils containing sufficient selenium can be toxic. When samples of the selenium-concentrating 'loco weed' that grows in some western American states have been burned, the selenium recovered has often been found sufficient

insulators where very heavy electric charges are involved, without being melted or destroyed, and without leaking-off the charge. This accounts for their use in high-quality spark plugs designed to operate at high temperatures. In addition, zirconia does not conduct heat very well, a characteristic that makes it an excellent heat insulator for high-temperature industrial processes. It can also be made into a mineral cloth that will persist indefinitely at a continuous temperature of 3000 degrees Fahrenheit—in fact, that will not melt until it reaches a temperature of 4700 degrees Fahrenheit.

Zirconia has many other uses. Its whiteness and opacity increase the covering power of high-

ABOVE: *Rocket-propellant studies showed these impact patterns made as burning aluminium particles spattered out of the fuel.* FACING, TOP: *Agena target docking vehicle.* RIGHT: *Shock waves during test of beryllium-coated nose cone.*

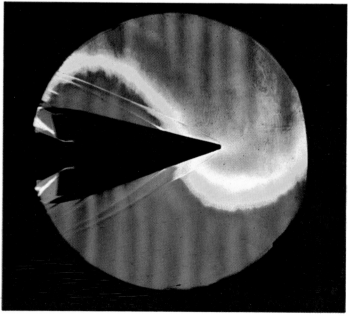

to poison grazing cattle. Of course, this poisonous characteristic can also be an asset. Selenium is used in insecticides and in special paints for ships' hulls to prevent the growth of algae, barnacles and other sea life.

The most unusual characteristic of selenium is its electrical behaviour. When selenium powder is heated it forms a glassy substance called vitreous selenium. This material is a good conductor of electricity as long as light is shining on it. In darkness, however, it resists the flow of current and is a poor conductor.

Many ingenious applications are made of this characteristic. For example, when the day grows

213

dim a strategically placed selenium cell—or electric eye—develops a resistance to an electric current flowing through it. Eventually as it grows darker the flow of current is reduced enough so that it flips an electric relay switch, and the street lights go on. A similar cell placed at the mouth of a factory chimney will record the fact that the smoke is too thick and the fires need adjusting.

In another application, a photographer points a selenium-cell meter at his subject. The meter records the changing flow of current from a tiny battery. This tells the photographer how much light there is and what his camera exposure should be. For the long-distance transmission of photographs, a selenium cell is run back and forth over a lighted photograph. It then sends corresponding electrical signals over a wire or by a radio broadcast. These signals control a small light source on the other end. This light, travelling over a sensitive photographic plate, will be bright or dark depending on the signal it gets from the selenium cell, and will consequently print an exact long-distance copy of the picture.

Selenium has non-electrical uses as well. It produces in glass the bright red colour needed for danger signals and traffic lights. When added to steels in small amounts it helps improve their machining characteristics for cleaner and easier cutting. Added to lubricating oils, it delays rapid aging and prevents gumminess. Perhaps oddest of all, selenium metal is usually bright and odourless, but when heated and vaporized it becomes putrid. Added to other poisonous gases, like industrial carbon monoxide, it is utilized as a warning of the presence of a poisonous gas leak.

Selenium mineral species are few in number and highly scattered; almost all selenium must be recovered indirectly. Selenium is reclaimed, for example, as a by-product of the copper refineries of the United States and Canada. There are also certain black shales in Idaho that contain as much as one pound per ton of ore.

GERMANIUM AND SILICON: In this electronic age, almost every synthetic and natural material is being investigated to see if it has some kind of electrical property that might be of practical value. Research has opened up a whole new field of practical electronics based on the semiconductors.

Some substances can move their electrons about with great ease—in other words, they conduct electricity very well. Some do not, because their electrons are too rigidly held in place.

Between these conductivity extremes are several elements called semiconductors. Normally they do not conduct an electric current, but with a little help they can be encouraged to perform. This help comes in the form of a small amount of impurity added to certain of these elements as their crystals are being formed. The process is called 'doping'.

Two of the semiconductors most frequently used are germanium and silicon. These elements look like bright silvery metals, but behave toward most other elements, in their chemical combinations, as if they were not metals. Several elements can be used for doping them. Aluminium, arsenic and boron are typical and effective. The kind of doping determines what kind of semiconductor will result; there are n-type (negative) and p-type (positive).

The major difference between the two types is that they permit what little current flow there is to move in opposite directions. When doped with arsenic, germanium metal is the n-type, and when doped with aluminium it is the p-type. Silicon doped with boron is p-type, with arsenic, n-type.

The most important use for these materials is to replace vacuum valves in electronic equipment. These semiconductor units consist of doped single crystals. Because of their size and the way they operate, they do not need extravagant amounts of power to heat them up, as valves do. They also last longer and can stand much more punishment than a delicate, assembled valve, with its many internal parts and glass or metal enclosure.

Best of all, semiconductor units can be very tiny and still very effective. They make possible electronic hearing aids that fit completely inside the ear and pocket radios as small as cigarette packages. Modern electronic computers can contain great numbers of tiny semiconductors, increasing their present use and potential.

Space technology has developed new minerals, new uses.
FACING, TOP: *Astronaut Edward White's space walk during Gemini 4's 1965 flight.* RIGHT: *Fluorescent dye marks recovery spot as U.S.S. Guam picks up Gemini 11 after its three-day earth orbit in 1966.*

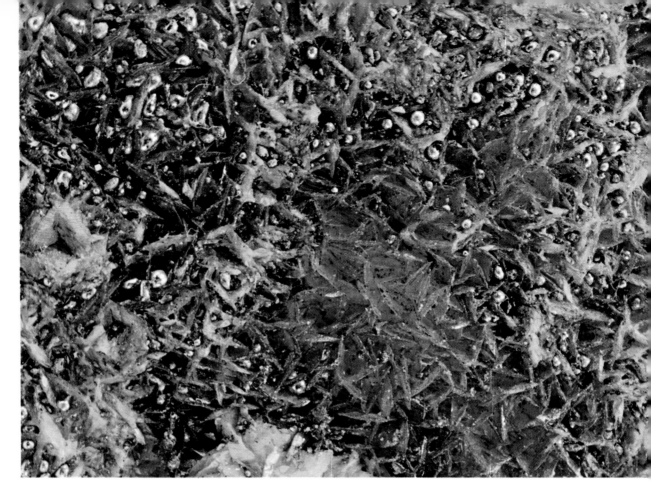

A group of unusually fine
specimens of ore minerals.
TOP: *Droplets of native mercury
from New Almaden, California.*
ABOVE, LEFT: *Proustite, a
ruby silver mineral, from Chile.*
ABOVE, RIGHT: *Dendrites of gold, Grass
Valley, California.*
RIGHT: *Chalcotrichite, in whisker
crystals from Morenci, Arizona.*
FACING: *A silicon integrated
circuit-chip which will pass through
the eye of an ordinary needle. It can
perform the functions of a conventional
circuit hundreds of times this size.*

The two basic jobs given to semiconductors that replace vacuum valves are as rectifiers and as amplifiers. Almost all electronic devices use rectifiers, which allow current to flow in one direction only. Arranged with certain combinations, semiconductors perform this function. In other combinations they become the now familiar transistors, which amplify weak current into strong and take the place of the amplifier valves formerly essential to most electronic gadgetry.

ANTIMONY: Antimony is one of several elements that sometimes behave as metals and sometimes do not. Arsenic, selenium, tellurium, chromium, tin, manganese, vanadium, tungsten and molybdenum are in this group. Antimony is used to manufacture 'hard lead'. Pure lead metal is too soft and flexible for most purposes, so from 4 to 12 per cent antimony metal is mixed with it to make it harder. This lead is used to manufacture many items, including car batteries, sheet and pipe for the chemical industry, and protective sheathing for telephone cables. If the percentage of antimony is kept low, thin lead sheet keeps much of its flexibility for such uses as the manufacture of collapsible toothpaste tubes.

Some of the alloys containing antimony are useful because this element leaves them with a smooth surface after casting. Also, these alloys expand slightly on solidifying, so that they can fit tightly to a mold and take on its finest details. Type metal, for linotype and monotype machines, contains 10 to 15 per cent antimony. This makes for smooth well-detailed type and also enables the alloy to melt more easily.

Like most metals, antimony and its compounds have been applied to military uses. Added to lead metal, it hardens shrapnel and bullet cores. The oxide of antimony prevents materials from burning and is used as a fireproofing spray for canvas. As an ingredient in paint mixtures, it makes a non-burning paint for ships. It has also been found that antimony sulphide reflects infrared light just as green tree foliage does. It makes an excellent camouflage paint against spying infrared devices.

Many familiar every objects contain antimony. Safety matches have about 3 per cent antimony in the match head and 8 per cent on the striker. Many new plastic floor tiles contain antimony oxide. Powdered antimony burns with a bluish flame when heated in air—a characteristic that can be seen in any fireworks display.

Stibnite (Sb_2S_3) is the most important ore of antimony, although it is also mined as oxides. Bolivia and Mexico are its leading producers.

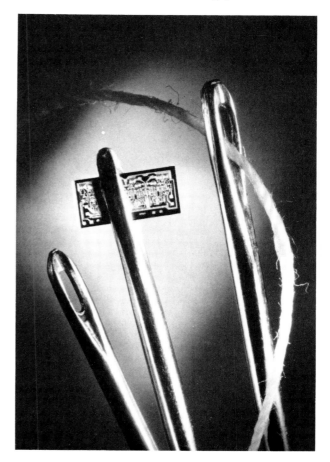

By far the greatest number of people involved, in one manner or another, with the mineral kingdom have had no professional training in the subject. More to the point, they have no desire to be professionals. They are interested in the earth sciences as a change of pace from everyday routine, or perhaps as a source of absorbing mental activity.

If this chapter seems to the reader to have a strongly American emphasis, it is because the amateur mineralogical scene in the U.S. is the most active in the world and shows how this fast-growing hobby is likely to develop in Britain, Australia and elsewhere.

Recent estimates put the number of organized hobbyists in the United States at over seventy-five thousand. The largest of the several American earth science hobby magazines, *The Lapidary Journal*, has a monthly circulation of over forty thousand copies. Since most people making collections or working with rocks and minerals are not in organized groups, it is reasonable to assume that they total several times this number in the United States alone.

The volume of sales of supplies, equipment, specimens and books has reached the status of big business in the U.S. In contrast with the situation ten or fifteen years ago, it is now possible for a dealer in these materials to make a comfortable living. Hobby dealers catering to the trade find it a full-time commercial activity. There is at least one trade journal, *The Lapidary Reporter*, that keeps the dealers abreast of sources of material, the state of the market, trends in the hobby, and centres of marketing activity. The total number of known dealers in the United States hovers somewhere around three thousand.

In Victorian times and a little earlier it was very fashionable in Great Britain to collect minerals and natural history specimens generally, and there were

many dealers to supply the demand. Interest declined until around 1960 when there were only about three mineral dealers in the whole country. The heyday of British mining is long since past, but the number of hobbyists is growing again and an interest in lapidary work is spreading from the United States.

A hobbyist may be completely casual—but surprisingly often he slips over the boundary into the status of a scientific professional. Some amateurs develop talents, not common among professionals, for rapid sight recognition of minerals, and can quickly identify several hundred species. The amount of time the hobby absorbs can vary from a little to a lot. There are many amateur collectors who spend every weekend and every holiday somewhere in the countryside searching for new samples of rocks and minerals. Families often plan their holiday trips around the locations of well-known rock and mineral exposures, and children often develop a rock-collecting hobby of their own through helping to add to the collection of one or both of their parents.

As with any hobby, the amount of money spent is not a direct measure of the enjoyment derived. There are hobbyists who spend a great deal of money on travelling and buying specimens. There are others who spend practically nothing; they prefer those specimens they can find for themselves and make only a minor investment in equipment. Still others, as a matter of principle, will not purchase any specimen material at all but rely on the shrewd exchanging of self-collected material to build their investment in the hobby.

Since mineralogy deals with objects and their manipulation, participation usually takes the form of making collections. A collection can be very general or very specialized. Some people collect gems, either cutting them at home or buying them already cut from dealers. Others collect

PRECEDING PAGES: *The amateur who starts with an interesting rock from his garden often ends up with the pick, maps and safety equipment of the expert field worker —and a growing home laboratory as well.*

mineral specimens. Of these, some are 'Dana' collectors whose collections include as many as possible of the species that are listed in Dana's *System of Mineralogy*. Collections are also made of minerals from a single mine or quarry or region.

Some collectors prefer single, isolated, well-formed crystals or pseudomorphs or particular chemical groups, such as sulphides or oxides, or even specimens remarkable for their resemblance to sculpture. In the 1870s at Lancaster, Pennsylvania, one Reverend Rakestraw became interested in tiny mineral specimens from the iron mines at Cornwall, Pennsylvania. These required magnification to be seen properly, and for their protection he began mounting them in small boxes. He would meet periodically with fellow collectors in Philadelphia, who soon picked up his interest. Unwittingly he triggered a chain of events leading to a cult of several hundred collectors called *micromounters*. Aside from the striking beauty of the micromounts, many minerals are simply not available in crystals of larger size and are ideally suited for such collections. Mineral collecting and collections are apparently limited only by the imagination.

Collecting, however, is only one of many areas of activity open to amateurs. They have found that a number of minerals are suitable for cutting and polishing into gemstones. Most of the easily available material is suitable only for non-commercial objects. However, many kinds of agate and jasper and common rocks are beautiful when cut and polished. Several of the colours and patterns they display are even superior to those of classic decorative minerals that have been cut for centuries and can be creditably exhibited beside such specimens or even in sculpture collections. In an effort to create minor—or even major—treasures, many thousands of amateurs have mastered the lapidary

A passing specimen collector pauses in Clarks Fork Canyon, Wyoming, and by contrast points up the size of an eroded pillar of red sedimentary rock.

arts. They are turning out exquisite creations, many equal to those produced by European commercial lapidaries during their best years. This craft satisfies the urge for manual work and at the same time provides an outlet for creative, artistic expression.

One of the more recent developments among hobbyists has been the bibliophile group—hobbyists who collect old books and manuscripts concerned with gems and minerals. There are also writers, lecturers, teachers and a variety of entrepreneurs who have been attracted into the field. Some have no particular mineralogical knowledge, but are interested enough to enjoy mixing socially with those who do.

Amateur mineralogists have a standing not common to amateurs in other fields. They have sometimes led the professional scientists to pre-

viously unknown mineral occurrences; they have helped find new species and provide a constant flow of new study material. As an example, many years ago in Colorado Springs, U.S.A., a mineral dealer and student of crystallography was intrigued by some crystals he found in material from the mines in Franklin, New Jersey. Familiar though he was with minerals, he could not identify these, and in 1911 brought them to the attention of Dr. Charles Palache at Harvard University. Eventually they were proved to be a newly discovered species. The description was published under the name cahnite after Mr. Lazard Cahn, the discoverer. His achievement was also commemorated on his tombstone, which—by his own direction—was made in the shape of a cahnite crystal.

The amateurs who become interested in the more technical aspects of mineralogy and geology frequently become amateur scientists, even if not prepared to go the whole way into professionalism. Often they expand into the related fields of chemistry, physics or mathematics. Even as a diversion, mineralogy offers a purposeful and mentally challenging leisure-time activity—as can be attested to by the professional scientists who began as hobbyists only to find that part-time involvement with this fascinating field was not enough.

The interested amateur soon discovers that he can pursue a variety of activities ranging from entirely physical to entirely mental work. If he decides to build a mineral collection, he will probably begin a succession of field trips, returning home to find that his work has just begun. Collections build slowly. They consist of the best specimens available at the time from all sources. They must be properly cared for and upgraded when the opportunity arises. This upgrading is a chore in itself; specimens must be cleaned, labelled, cata-

FACING, TOP: *Radioactive torbernite from Aveyron, France—unavailable to collectiors.* LEFT: *Crocoite from Tasmania.* ABOVE: *Dolomite from Germany. Pieces similar to these two may still be obtained.*

223

logued, properly arranged and stored. Some must be trimmed or repaired; others must be sawn open and polished on exposed surfaces. Specimen preparation has become a fine art and a little appropriate trimming often doubles or triples the quality and the value of the piece.

Aside from these basic curatorial chores, the collector may want to prepare a display cabinet or a fluorescent exhibit or a series of special groupings highlighting perhaps radioactive species or pseudomorphs. He may become excited about competing in shows; assembling a competitive exhibit is painstaking work. A serious collector, of course, will spend time studying the minerals themselves to enlarge his knowledge.

The collector, in fact, may be much more interested in studying minerals than in collecting them. In this case, he need not enter the mad scramble of competing for high-quality or expensive specimens. He may collect his own, rely on gifts, or purchase less desirable and massive pieces. At home he will have a small laboratory which can become impressive in time. It will probably consist of a mineral table, mineral corner, or mineral alcove, unless he is starting out grandly with a whole mineral room. If properly set up, all his materials, including books, will be handy in one place. There will be a book of mineral tables, a set of hardness-testing minerals or test points, a magnifying glass, a streak plate (a rough-surfaced porcelain tile against which the specimen is scratched to reveal the colour of the powdered mineral), a magnet and an ultra-violet light. Later, more expensive equipment will be added—a microscope, a refractometer, and either a manufactured or homemade specific gravity balance. From there on, as he develops competence, other specialized equipment may be added, including a blowpipe and chemical testing equipment.

When the hobbyist's main interest is in art and craft work, he will find several excellent texts to instruct him. For centuries, professional lapidaries passed the secrets of the trade from generation to generation, keeping close control over their knowledge of methods. Gradually, however, the secrets were worked out by amateurs and published, and inventive minds turned toward improving and simplifying the old techniques. Now the beginner is enriched—or plagued—by a bewildering array of tools, machinery and advice for all sorts of lapidary endeavours.

It is fascinating for the aficionado to try out new ideas, just as it is for a housewife to try the delicacies in a new cookery book. To start with, he will want a saw for trimming off pieces of rough material to be worked into gems, a grinding wheel for starting cabochons, a sander for fine grinding, and a polishing buff. These may come in combined

FACING: *Gold panning at Uralla, New South Wales.*
TOP: *Engineer Paul Seel is an amateur who has become an authority on diamond crystal forms.* ABOVE: *An amateur expertly cuts and grinds a cabochon.*

units with a single motor, or they may be bought separately. Later, faceting equipment may be added. By this time he will be familiar enough with the books and with other lapidaries and dealers to know what further equipment he needs. By this time, also, he will have decided whether his talents lie in making cabochons, faceted stones, carvings or other lapidary objects. He will know what to look for at sales and what to deal for when trading with other collectors.

Meanwhile he will probably have established contact with other hobbyists who are proficient in precious-metal casting and shaping. A number of amateurs drawn together in this manner have been functioning as new, creative jewellery designers, supplying unique pieces to a growing market. The demand for handcrafted jewellery has enlarged greatly since so many fine materials and new ideas have started to supply it.

Ultimately, all materials used by the hobbyist, except for manufactured equipment and supplies, must come out of the ground. It is surprising how much of it is excavated through amateur activity rather than by professional suppliers for the trade. Every fresh opening in the earth, especially when it cuts through rock, is a potential source. Every mine, mine dump, quarry, stream bed, tunnel or building excavation becomes a mecca for collectors. It is almost certain that every new excavation in the last few years has been visited by at least one rummaging hobbyist. Vast tonnages of rock are hauled home to be worked over, reduced in volume and carefully studied; from these treasure troves come many specimens for collections or future lapidary work or even for sale.

The problem of accommodating numbers of visiting collectors is often a nuisance not only to mine and quarry operators but also to any property owner with an interesting outcrop on his land.

Many instances are known of property damage and personal injury arising from the activities of careless 'rockhounds', and owners may show their lack of sympathy by declaring the properties closed to collectors and posting guards to show they mean it. Nevertheless, there are ample collecting sites and new ones are found constantly by questing enthusiasts. The collector should always take care to respect property rights, and seek the owner's permission before collecting. Occasionally one may find a pile of hand-picked ore near an old mine dump, and should remember that it may well represent the hard work and livelihood of a lone prospector. In Britain there are no strictly applied laws relating to mineral collecting, but in Australia each state issues a publication on the legal rights of collectors which apply to amateur and prospector alike. It may be necessary to obtain a permit, which should not be overlooked since the increase in number of amateurs has caused the courts to take infringements seriously.

Planning for collecting trips often takes on the aspects of a full-scale scientific expedition, involving technical briefings, maps, travel directions, and camping arrangements. On the other hand, the expedition can be as simple as a Sunday afternoon trip in the family car to a rock outcrop in the country. Often the trips are taken in groups sponsored by the local mineral society. But to minimize nuisance and maintain goodwill, numbers should be kept down and visits made not too frequently. There are also highly popular professional tours, led by experienced guides, to exotic collecting grounds in Mexico and Brazil.

The collector's field equipment is usually rather simple. It may consist of a rugged, geologists' hammer, rock chisels, wrapping paper for specimens, a magnifier for small crystals, a carrying bag, a small

notebook, and a first-aid kit. A surprising number of cars travelling on the roads these days keep such simple equipment in the boot, ready for any opportunity. Of course, there are always some collectors who go equipped for certain special mining projects. It is not unusual in the western United States to come across rockhounds working over an outcropping with dynamite and mining tools, and there are many books to teach them the techniques. Particular care must be taken to observe laws relating to the use of explosives.

Once the material is out of the ground, the collector saves some of it for immediate or potential use. The rest finds its way into circulation. Depending on the whim of the finder, it is sometimes dispensed as

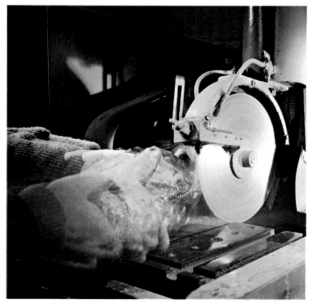

TOP: *Pyromorphite from Ems, Germany. Specimens from this old occurrence are still sometimes available.*
ABOVE: *Basic equipment for the amateur laboratory is an inexpensive trimming saw, which is fairly easy to master.*

gifts to collectors, schools or museums, exchanged with other collectors, or even sold to specimen dealers for income or for money to purchase other specimens. Specimen gifts made to friends or children have attracted many new converts to the hobby, and collectors often give away specimens with this object in mind.

Swapping specimens is as old as commerce itself. A shrewd mineral trader can improve his collection substantially if he has the patience and perseverance to carry out the negotiations. Much of this exchanging takes place by post. All sorts of ingenious packing techniques have been devised by collectors to prevent destruction of specimens in transit. Nesting a tissue-wrapped specimen in a box packed tight with popcorn, or burying the specimen in a box of soap powder, which can be washed off on arrival, are ingenious examples.

There are lazier methods of building a collection. The quickest, surest and easiest way—particularly to acquire special specimens—is to purchase from a dealer. Roadsides in the western United States are peppered with rock shops catering to transient collectors. Some dealers have set up respectable mail-order business contacts through advertising. Others set up sales booths at rock and mineral shows. Through these, they sell enough to pay expenses, gain a small profit, and at the same time establish new contacts with collectors who become steady customers.

Collectors in America have long referred to the custom of buying specimens as 'silver-picking'. Highgrading—not in the pejorative sense of the mining fields—is another of the terms in the hobby language. It means that the collector has gone over a dealer's stock or a mine dump or a collection being sold and has taken out the best things, leaving the rest for others. Arranging somehow to get at a

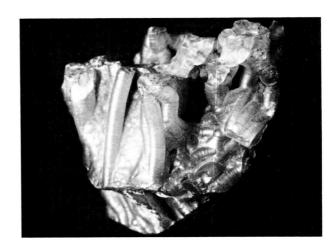

dealer's new stock and highgrading it before it goes on general sale is a coup every collector dreams about. One dealer, in an attempt to avoid partiality toward any of his customers, used to advertise each year a special sale at a New York hotel. The specimens were laid out on tables, but no one had access to them until the hour of the sale. Several hours before, collectors would begin to queue up outside. At the given moment the doors were thrown open and the stampede began, with collectors grabbing specimens from the tables and from each other. This kind of special sale was eventually abandoned, with a return to the less hectic, first-come-first-served procedure.

TOP: *Australian chalcedony.* ABOVE: *Collecting agate in Agate Creek Valley, Australia.* FACING, TOP: *Gold washing in a mountain stream in the United States.* RIGHT: *Sieving rubies.* FAR RIGHT: *Testing for fluorescence in Franklin limestone.*

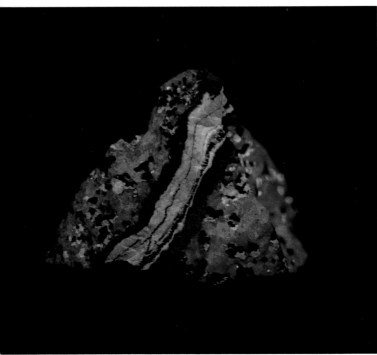

The amount of money being spent on specimens has now reached such proportions that there is tremendous incentive for dealers to obtain more and more material. This in turn has stimulated production on down the line. More specimens are now flowing into the market than at any time before in history. The demand is so great that some mines, not good enough for commercial ore production, are being operated on a small scale just to produce specimens.

The pressure on the market has also stimulated a creeping inflation of prices. In turn, this has accelerated the rate of supply, because newly mined specimens bring better and better prices. In an effort to eliminate the middleman and his profit, the dealer has tried to get closer to the sources of supply; this has brought higher returns directly to the miners. As an example of the increased activity, years ago only a handful of wholesalers worked regularly through certain active mining areas in Chihuahua, Mexico. These same areas now play host to crowds of buyers each year.

Various sources of information are available to help the collector prepare for field excursions, discover what other collectors are doing, and keep in touch with the dealers. The widely circulated American *Lapidary Journal* publishes an extra thick issue every April called the 'Rockhound Buyer's Guide'. In this hundreds of dealers are listed, cross-indexed to a list of their specialities. This issue also carries pages of advertisements for merchandise of interest to the hobbyist. Everything, from metal jewellery parts or findings to complete camping outfits, is likely to be represented. Reference books are listed, including range guides for particular localities, textbooks, and how-to-do-it books for the lapidary.

Although the United States mineral addict has many publications devoted to his needs, the British

and Australian enthusiast is not so well served and has to rely more on textbooks and the monographs and handbooks issued by national and state geological surveys. Other usable information may be obtained from fellow collectors, museums, university departments of geology, highway and mines departments, and local chambers of commerce. The avid hobbyist soon becomes expert at quarrying the information he needs from every source available.

The amateur in the United States who wishes to join with others who share his dedication has plenty of opportunity. Across the country, there are more than seven hundred amateur societies open to him, varying in size and in specialization but all offering contact with other eager hobbyists. Most of the groups include 'Gem and Mineral' in their titles. Some serve a broad range of interests, while others vary from mineral study groups to practical lapidary clubs. In general the western groups, particularly in California, tend to be larger. Some of the societies are sponsored by institutions, some are self-sponsored. The hobby magazines are the best source for all sorts of club information. They can guide the hobbyist to a club in his area or advise him on how to start one if none exists. Through them, as well as through local newspapers, the hobbyist can keep track of nearby mineral and gem shows, an excellent way of making contact with others interested in the field. The most useful American magazines are: *The Lapidary Journal*, Box 2369, San Diego, California, 92112; *Rocks and Minerals*, Box 29, Peekskill, New York, 10566; *Earth Science*, Box 550, Downers Grove, Illinois, 60515; *Gems and Minerals*, Gemac Corporation, Box 687, Mentone, California, 92359.

The principal Australian journal is the *Australian Lapidary Magazine*, now in its fifth year of publication. In its first three months its circulation rose from 750 to 5000 copies.

Specimens as fine as these can now be found in private high-quality collections. FACING, TOP: *Legrandite from Mapimi, Mexico.* LEFT: *Fluorescing willemite from Franklin.* ABOVE: *Wulfenite from the Glove mine, Arizona.*

Membership in a well-organized group can greatly enrich the hobbyist's experience. An active club always includes at least a few field trips on its calendar, as well as indoor lectures and demonstrations. Popular and well-informed lecturers move from group to group, heralded by word of mouth and through the magazines. Members are continually improving their information through panel discussions, slide shows, and small serious study groups, and improving their collections through swap sessions and by attending specimen auctions.

In the United States the numerous earth science and lapidary societies have been organized into six regional federations. Each federation has its own set of officers elected by member clubs. In general, the duties of the regional federations are to help the member societies to thrive. They distribute programming aids and information, conduct scholarship drives, and organize the annual convention and and show held in each region.

The regional show is usually the big event of the year. It will typically be held in a large convention space. In the centre area, many rows of showcases display competitive exhibits entered by individuals and clubs, covering various aspects of the hobby. A set of uniform rules for competition is used throughout the country, setting up nationwide standards for judging. The exhibitor, then, is actually competing with other collectors on a nationwide basis.

The dealer's booths—the business part of the show—generally surround the display area. Elbow to elbow, customers and curious onlookers crowd around to look at the latest materials and to window-shop before making final purchases.

By the time the average collector leaves the show, he has gorged himself on seeing displays, meeting friends, attending nearby field trips, and swapping yarns and specimens. He may also have spent all his money and been reduced to making mental notes of things he would like to buy next time.

The regional federations themselves are organized into the American Federation of Mineralogical Societies. It assists the regional federations in various ways, maintains the uniform judging rules,

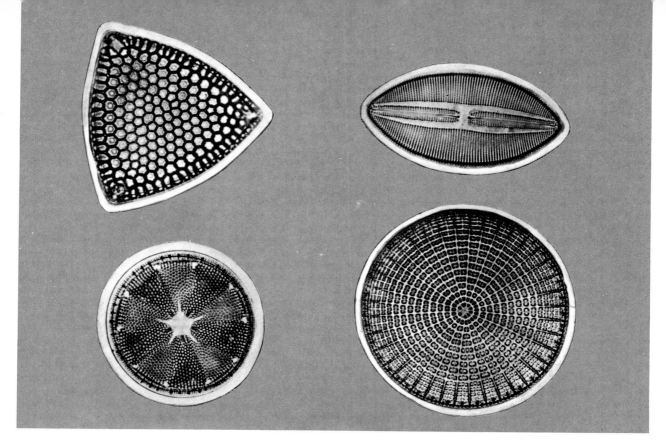

and in general oversees the hobby on a national scale. It also sponsors the most important show of the year. Top awards at these shows are considered the pinnacle of success. A competitor must win a first place in a regional show before he is even allowed to enter into national competition.

Since the show—whether local, regional, or national—is such a vital centre of interest, it is not surprising to find a number of enthusiasts travelling from show to show. To accommodate them—and also the various federation officials—all the big shows carefully schedule their dates to avoid overlap.

British clubs are few, and although considerable increase in their number is likely, collecting possibilities are limited. A combination of high density of population and a long history of mining and prospecting makes exciting new finds of good specimens improbable. Australia, on the other hand, despite its known riches of mineral resources including gemstones and recent prospecting booms, remains largely unexplored in detail and the future

is rosy indeed. The founding of the Gemmological Association of Australia in the late 1940s, followed by the Lapidary Club of New South Wales, created widespread interest; gem clubs and rock shops mushroomed in every large town and city. The first frenzy has evaporated somewhat, but steady growth seems inevitable. American experience shows a healthy example of the way that amateur collecting activities can be organized and encouraged, and the goals are well expressed in the constitution of the American Federation: '. . . To promote popular interest and education in the various earth sciences and in particular the subjects of geology, mineralogy, paleontology, lapidary, and other related subjects, and to sponsor and provide means of coordinating the work and efforts of all persons and groups interested therein.'

FACING, FAR LEFT: *Copper dendrite from Bolivia.*
LEFT: *From a Virginia quarry, apophyllite on prehnite.*
ABOVE: *Diatoms (1000 times enlarged) have plant and animal characteristics, and persistent silica skeletons; they are important rock builders.*

The Connoisseur
10

According to Webster, an amateur is 'one who cultivates a particular pursuit, study, or science, from taste, without pursuing it professionally.' A connoisseur is 'one aesthetically versed in any subject . . . one competent to act as a critical judge of an art, or in a matter of taste.'

The amateur, then, can always aspire to connoisseur status—the amateur who can afford the time, who is a true devotee, and who is equipped with or can develop great patience and perseverance. From such collectors come the few who rank as connoisseurs.

The connoisseur's collection eventually abounds in specimens of exceptional quality, beauty, rarity or scientific importance. He has accomplished this by refining his appreciation of beauty and expanding his knowledge of minerals. He has had to learn about mineral uses, phases, varieties, relationships to rock, relationships to each other, associations, intergrowths, differences of locality habit and other scientific data.

Few collectors, even among connoisseurs, can afford to be universally well prepared in all these areas. Generally attention must be focused on some special excellence—the physical perfection of the specimens, perhaps, or the completeness of the collection. There are no restrictions, however, and one may specialize or be general as one pleases. Some of the greatest mineral collectors have collected both scientifically and aesthetically, with equal success in both areas.

Whenever a great private collection is mentioned, the collector is spoken of as a man of high aesthetic sense. Aesthetic appreciation is very elusive and difficult to define. Some people just seem to have it and some do not. One certainty is that anyone with this ability is able to compare various objects and to discriminate among them, seeing differences that the less well trained observer might overlook.

In order to place a value of any kind on any object it is necessary to have something of known value for comparison. Anyone making an evaluation of the beauty of a mineral specimen is in a much better position if he has had broad experience with other mineral specimens. After all, a cat will never win any beauty prizes at a dog show, especially with all the judges knowing only dogs. Every kind of object has its own standards of beauty, which must be thoroughly understood if they are to be applied with any consistency or intelligence.

There is some overlap of aesthetic appeal, of course, among various kinds of objects. Among mineral specimens certain forms can be reminiscent of sculpture, and certain colour combinations reminiscent of different schools of art. It is possible, then, for collectors with a general appreciation of the arts to assemble collections that are aesthetically attractive. Because this kind of aesthetic judgment is the easiest and most familiar path toward specialized collecting, many more collections are founded on aesthetically pleasing specimens than are based on scientifically significant ones.

Eventually, the connoisseur faces the critical problem of trying to find more specimens of the highest beauty. There never have been, and are not now, enough of them to satisfy demand. As a result, severe competition for a few specimens continually drives prices higher, making the collector's income the primary instrument of his success in building this connoisseur's kind of collection.

Connoisseurs of mineral specimens can be distinguished from ordinary collectors by the extent of their knowledge of the subject. If there is any common measure for this group, it is this combination of knowledge and intelligence. By contact with each other and with the great museums, these mineral

PRECEDING PAGES: *A collector becomes a connoisseur when he is able to recognize the best of everything that relates to his hobby, for only then can he begin to acquire it.*

collectors have established certain standards for specimen selection: beauty, rarity, perfection of form, scientific importance, and illustration of some natural phenomenon. Specimens illustrated in this book are entirely within this tradition and have been selected for precisely these reasons. Perhaps the best way to develop a feeling for these criteria is through frequent visits to any of the good museums. Almost all of the important institutional collections of the world are founded on private collections assembled by connoisseurs.

Well-established and securely founded museums are a collector's key to the future. They offer a centrally placed, accessible, properly equipped repository for the preservation of a collection that may have taken a lifetime to build. Such a collection becomes important to posterity. Its dispersion and destruction would be a great waste. The best museums maintain and increase their collections much as the private collector would have done, thus continuing his life's work. They have show collections on public exhibition and study collections in drawers, as most private collectors do. They contact the sources of new specimens and try to acquire them, just as the collector did. By maintaining and improving the collection, they make the original collector's efforts even more worthwhile.

The pattern of great institutional collections growing round a nucleus of private collections holds true around the world. The famed British Museum, repository for many private collections,

Many extraordinary amethysts have come from Guerrero, Mexico, an isolated spot from which the finds must be carefully packed out on donkey-back. Largest and finest of its kind is the two-foot-high specimen above.

was founded on that of Sir Hans Sloane in 1753. This collection was purchased, as were many of the other collections acquired in the first century or so of the Museum's climb to its present eminence; of special note are those of the Right Hon. Charles F. Greville (acquired 1810), Thomas Allan and R. P. Greg (1860), and General N. I. Koksharov (1865). Each of these collections contained many specimens described or illustrated in important reference books on mineralogy. In the 1920s and '30s, F. N. Ashcroft presented a magnificent collection of Swiss minerals with a very detailed catalogue, which has become a standard reference collection. Even more recently, in 1964, the Museum received by bequest Sir Arthur Russell's collection of some 14,000 specimens of British minerals, probably the finest regional collection ever made.

Of American museum collections, three are by far the finest and are among the handful of first-rank collections of the world. They are the collections at the American Museum of Natural History in New York, the Department of Mineralogy of Harvard University in Cambridge, Massachusetts, and the Smithsonian Institution in Washington, D.C. These three have reached their present state of major importance among collections throughout the world because of the prodigious collecting activities of a comparatively few private collectors.

The tradition of fine private collections of minerals began in the United States around 1790 when a botanist, David Hosack, brought a collection from Europe. Eventually it came into the hands of Princeton University, where it became the first public exhibit of minerals. Later its identity was lost among other collections at Princeton.

The early American collectors were all located on the East Coast which, at the time, was the financial, educational and cultural centre of America. It was not surprising, then, that when the first golden age of American mineral collectors arrived at the end of the century, it was concentrated in the area from Philadelphia northward. Prominent American collectors today sometimes speak of mineral collecting 'in the old Philadelphia tradition'.

The group of collecting titans of this period included Washington A. Roebling, Frederick A. Canfield, Clarence S. Bement, George Vaux, Magnus Vonsen and A. F. Holden. Each of these men assembled a collection sufficient in itself for an entire museum. Perhaps the most important private collection that was ever assembled in the United States belonged to Washington A. Roebling (1837-1926).

Colonel Roebling was an engineer who is famous for building the Brooklyn Bridge and for his mineral collection. He was made an invalid by an attack of the bends, which occurred while he

*In any field, masterpieces are established by the taste,
experience and trained eye of the connoisseur.* FACING:
Amethyst from Due West, South Carolina. ABOVE: *Jan
Vermeer's 'Woman Weighing Gold'.*

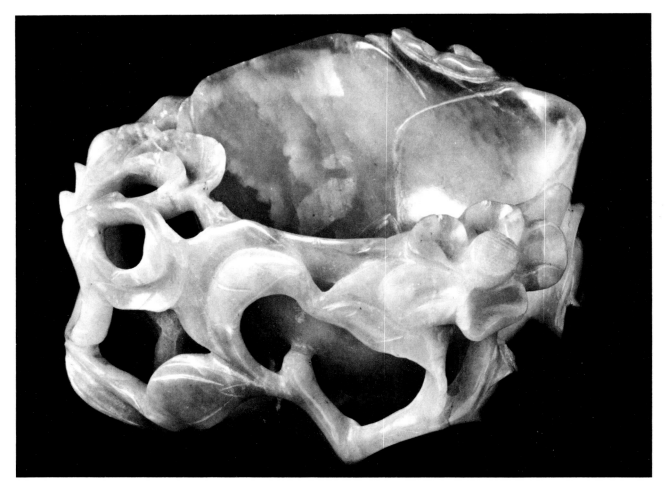

was working in the caissons under the East River during construction of the Brooklyn Bridge. Because of this, he did much of his collecting by correspondence from his home in New Jersey. With sufficient wealth and ample time to work at it, he set very elaborate and worthwhile goals for his collecting activities. It was his intention to collect a sample of every well-described mineral species then known to man. By the time of his death he had come within a dozen specimens of his goal. In addition, he had gathered together hundreds of examples of varieties of these species from thousands of different occurrences around the world. Many of these specimens were of extraordinary beauty, unique in size or form or some other characteristic.

In the process of collecting, the Colonel became extremely knowledgeable about minerals and skilled at mineral identification. He made a point of offering his specimens for the use of mineral scientists of the day and knew most of these men personally. When he died in 1926 his son, John Roebling, gave the collection of sixteen thousand specimens to the Smithsonian Institute, which already had a good, but not superior, collection. Overnight, with this gift, the Institution's collection gained rank with the leaders of the world.

That same year the Smithsonian received the Canfield collection. This collection's history began in the early 1800s when Mahlon D. Dickerson began to assemble a fine collection of minerals at his home in Dover, New Jersey. These were from various world occurrences, but luckily included some of the highly interesting mineral specimens being found at the time in New Jersey.

On Mr. Dickerson's death the collection went to his nephew, Frederick Canfield Sr., who built it into a fine collection of sixteen hundred pieces. Many of these again were from the famous zinc mines at Franklin, New Jersey. Mr. Canfield's professional connection with the mines gave him access to this material and he assembled what remains the finest collection made from these mines. On his death, the collection passed to his son, Frederick A. Canfield Jr., also a mineral collector. Keeping the original collection intact, the new owner began to assemble one of his own, which he built to a total of nine thousand specimens. This was also a worldwide collection, but was very rich in newly recovered, fine specimens from New Jersey.

Canfield subsidized workmen in railway tunnels, stone quarries and road construction to collect specimens for him. Like Roebling, he became very knowledgeable about specimens in the process of collecting them. At his death, having no heirs, he left the entire collection to the Smithsonian, along with an endowment for its increase and improvement.

With the Roebling and Canfield collections arriving in the same year, the Smithsonian was barely able to cope. But by the time the collections were absorbed a magnificent nucleus had been established to which, as time went on, other private collections have been added. Without question, the Smithsonian's National Collection of Minerals has been the result of the initial work of amateurs and connoisseurs.

When A. F. Holden graduated from Harvard University in 1888, he was best known as a famous football player. But he had been trained as a mining engineer, and soon became wealthy at his profession. Meanwhile he had developed a love for minerals, which he started to collect seriously in 1895. In a

short time his collection was one of the best in the country. Like Roebling, Holden was not interested in mineral species only as showy specimens. Nevertheless his collection does contain hundreds of small but beautiful specimens as well as great numbers of species samples. In 1913 he left the collection, with an endowment, to Harvard University. This gave Harvard's collection the leading position it now holds among collections of mineral specimens to be found in American universities.

The other of the three major American institutional collections, at the American Museum of Natural History in New York, achieved its status by the same means. Most important among its collections is one assembled by Clarence S. Bement, a wealthy Philadelphia manufacturer who searched out beautiful and exceptional specimens of their

FACING: *Carved nephrite, a Sloane specimen at the British Museum (Natural History)*. TOP: *Sir Hans Sloane.*

241

kind. His collection did not contain the broad sweep of species that distinguished the Roebling and Holden collections. However, by the end of the 1800s it was rated the highest-quality collection in the United States and among the finest anywhere in the world. Losing interest about 1900, Bement sold his collection to J. Pierpont Morgan of New York, who presented it to the American Museum. Many of its fine old specimens are still the best of their kind in existence.

No listing of great private collectors would be complete without reference to the Vaux family of the Philadelphia area. William Vaux (1811-1882) was one of the earliest of this rising group of wealthy connoisseurs. In his time, his collection was considered the finest in America. It was not surpassed until the time of Bement. Upon his

death, the William Vaux collection was left to the Academy of Natural Sciences in Philadelphia, where it is now the backbone of the Academy's excellent collection.

Before the collection went to the Academy, Vaux's son, George Vaux, selected from it certain specimens for himself, as he had been instructed to do. This nucleus stimulated the heir into building a collection of his own, which soon became a major one. George Vaux Jr. continued the effort until the final collection surpassed that of his grandfather. It was given to Bryn Mawr College, automatically placing this college collection among the more important ones.

During this age of great collectors almost all the collections were being established in the eastern part of the country. However, in Petaluma, California, Magnus Vonsen (1879-1954) began assembling a major collection about the time of World War I. Like all the other collectors he had the means, the time and the interest necessary. Also, like the others, he became very knowledgeable in the process. It was his dream that eventually a great public collection would be established on the West Coast, and to further the idea he left his collection to the California Academy of Sciences, which displays it in a museum in Golden Gate Park in San Francisco.

Many excellent private collections are being made now in various parts of the world. Some will eventually go to museums as they have in the past. The Natural History Museum of Stockholm has the beautiful Sjögren collection in its handsome inlaid wood cases. The Natural History Museum in Paris has the magnificent collection of Colonel Louis Vesignié, who died in 1954. It is worth stopping to consider Colonel Vesignié as a prime example of the true connoisseur—a collector whose energy, taste and discrimination made him a spiritual descendant of those enlightened amateurs of the eighteenth and nineteenth centuries.

In 1956, the 'Bulletin de la Société française de Minéralogie et de Cristallographie' printed a memorial article on Colonel Vesignié, describing him as 'a passionate mineralogist'. It is translated here, in part, because it offers almost a blueprint for the attitudes and equipment of the top-rank connoisseur.

'[Colonel Vesignié's] collection of crystallized minerals is among the most beautiful amateur collections in the world. At an estimate, it contains about 40,000 specimens. . . . His collection of precious stones is equally remarkable. It includes a series of pegmatite gems of which certain specimens are exceptional for their size and clarity; they are now displayed in the mineralogy gallery of the [Paris Natural History] Museum, where one can see along with other cut stones a 377-carat blue topaz from Siberia, a 250-carat rose beryl from Madagascar, and a superb alexandrite of about 100 carats. . . . He had brought together also a fine collection of meteorites, and he watched for the announcement of new falls to acquire good samples of them. . . . Colonel Vesignié sought after rare specimens, not only to please his eyes and for the satisfaction of being the sole owner of a unique specimen, but also for truly scientific reasons. In fact, from his conversation, it was easy to detect his perfect knowledge of mineral species, their modes of occurrence and association. This knowledge and these aptitudes are reflected in the composition of his collection, where natural series and paragenetic minerals are perfectly represented from the basic species to the rare ones.'

It is not only the wealthy who are able to participate in such endeavours. The fact that so many people of modest means are making fine collections is good evidence that wealth is not necessary. True, such collections do not become as large as those of affluent collectors, nor do they grow as quickly, but they can be every bit as good. Sometimes they command notice in their own right. Often, they are absorbed into other collections until, in aggregate, they form a single major collection. Such collecting can readily be done today and the specimens to make up such a collection are constantly available.

Collectors have no difficulty in finding good representative specimen material. It is the connoisseur who has the problem, because he must restrict himself to the rare and beautiful specimens that are so hard to get.

Several attempts have been made to develop a scientific method of appraisal as an aid in evaluating

FACING: *This seven-inch azurite crystal is one of the finest of many superb specimens from the copper mines at Tsumeb, South West Africa. Mining here has passed the oxidized zone that produced such crystals; no more are likely to emerge.*

the desirability of a mineral specimen. This has been done in the belief that value and desirability are closely related in our economic system. No attempt has been too serious and none completely successful. Judgment of fine specimens is still a cultivated art—one that takes considerable time and effort to develop.

George L. English, a noted mineral dealer, relates the story of the purchase by Clarence S. Bement of an excellent collection owned by Norman Spang of Etna, Pennsylvania, sometime around 1880. Each man made a personal appraisal of the collection in an attempt to arrive at a satisfactory agreement. Although the total was nearly $22,000, the two men were only a hundred dollars apart in their estimates. This indicates that there was a common set of values subtly accepted by both collectors. Other incidents of this kind occur frequently throughout the years, affirming that there are some unconsciously recognized factors controlling the desirability and therefore the value of specimens. George English suggests that these are commercial value, chemical composition, form, rarity, and a few other miscellaneous characteristics —beauty, size, hardness, uniqueness, and associated minerals. At first glance these seem reasonable factors for judging, but in every case the effect on specimen value is arguable.

For example, a thousand-pound crystal of beryl has considerable value if sold as an ore of beryllium. However, aside from its possible desirability to a museum or two, there is hardly anything less salable as a specimen for a collection than a crude thousand-pound crystal of any species. Its value, therefore, is only as much as the ore is worth. No collector would pay a premium price for it as a desirable specimen because its bulk, weight and roughness make it undesirable for a private collection. On the other hand, a one-ounce well-formed crystal of gold has a commercial value of exactly thirty-five American dollars as gold bullion, but carries a large premium value for collectors. Large gold crystals are rare and attractive. A one-inch, lustrous, twinned crystal of cumengeite has no commercial value whatsoever and yet among connoisseurs it is infinitely more desirable than gold and diamonds and brings equivalent prices. Commercial ore value, then, has little if any effect on the degree of desirability of a specimen.

George English also tried to make a case for the effect of chemical composition on specimen value. Here he increases the confusion by considering the premium value of rare chemical substances while also making allowances for commercial ore value. There is a good reason, however, for considering chemical composition as a determiner of value. Some substances are very rare. Newly described species particularly are in short supply. Collectors have difficulty getting them. This tends to heighten their desirability, even when they have little else to recommend them. The chemical composition in this case is only indirectly the cause; it is rarity that is the immediate consideration.

Specimen form certainly influences value. Every educated collector knows that specimens with large, perfect crystals and brilliant faces are the most valuable. This seems to apply regardless of the species. Species like quartz and calcite, even though very common, are generally much preferred to rarer or more unusual species that just do not occur in good crystals. The perfection and size of the crystals, and freedom from damage, distortion, coatings and inclusions, all have direct and strong bearing on value.

There are all sorts of variations on this theme. Specimens with crystals attractively scattered on

rock are usually better than those with crystals closely bunched. Specimens with a few large crystals are better than those with a druse of many small ones. When a species does not occur with crystals, any attractive form it may exhibit increases its value. Botryoidal hematite, stalactitic limonite, reticulated cerussite, coralloidal aragonite and globular pectolite are all more desirable than formless masses of the same minerals.

Rarity, like form, also seems to have a direct effect on the value of a specimen. Two kinds of rarity are generally recognized. One is the rarity of a specimen's high superiority. Very fine specimens, with all the requirements of crystal perfection, crystal arrangement, freedom from damage, and size of crystals, are exceedingly rare. The other kind of rarity is due to scarcity of the species or of a particular variety having unusual colour or form. Good crystals of garnets in dodecahedron form are common, but a well-formed cube of garnet—a legitimate crystallographic possibility—is rare indeed. George English, in his treatment of rarity, admits that when 'superlatively fine crystals of a mineral colour occur in great abundance' their value, based on the rarity of high quality, collapses.

All other miscellaneous factors determining price boil down to qualities contributing to the durability and beauty of the specimen. Beauty, in particular, seems to have a strong effect on specimen value, perhaps more than anything else. Admittedly,

Mines at Alston Moor, Frizington, and elsewhere in Cumberland, England, once produced many exceptional baryte specimens like the one above. An occasional piece may still be obtainable from a dealer or an old collection.

beauty is indefinable. Two specimens of the same mineral, almost identical in size, number of crystals, lustre, and perfection of condition, may differ greatly in beauty. One may have transparent crystals of soft pink colour and the other may be opaque and dull grey-green. Colour, shape, arrangement, transparency, neatness, and other subtle and related factors contribute to estimates of beauty. The collector with a cultured aesthetic taste will recognize the necessary combinations of these attributes at a glance.

By far the most overwhelming effect on specimen value is created by the operation of the laws of supply and demand. The mineral specimen market is now large enough that prices for desirable specimens are controlled by market conditions. Successful mineral dealers know their customers and their customers know each other. A balance is struck. The dealer will recover for his specimens what he knows the collectors will pay. Collectors on the other hand will pay what is necessary to keep all fine specimens from going into the hands of their competitors. The limit seems to be how badly the collector wants the specimen, based on his judgment of how rare or beautiful it is. This aspect of collecting does not lend itself to analysis; it remains an art. The most successful artists build the finest collections.

The true connoisseur, then, is not satisfied with simply accumulating specimens of great beauty, rarity, or perfection of crystal form. He also wants to know the life history of his specimens, the conditions under which they formed, and the natural transformations and processes they have been through. In his collecting activity he finds great peace and relief from the monotony and pressures of his business or profession, as well as an exciting mental tonic. He browses around mineral dealers'

displays, wanders in and out of their shops, visits auction houses, antique and second-hand furniture shops, other collections, museums, mines, quarries, excavations—any place that hints of a source of specimens. Through his hobby he is enjoying a fascinating cultural activity and at the same time he is increasing his knowledge of the mineral kingdom. And very likely—because of his deep acquaintance with the earth and its products—he understands the mineral foundation that is the basis of our industrial civilization.

Even museums must now compete
for specimens of this quality.
(a) Stibiconite replacing
stibnite, from San Luis Potosí,
Mexico. (b) Pyromorphite from
the Merkur mine, Ems, Germany.
(c) Mimetite from Cumberland,
England. (d) Proustite
from Chanarcillo, Chile.
FACING: Gold latrobe nugget.

a

b

c

d

Index

*Numbers in italic refer
to illustrations*

Credits

The publishers wish to thank the Smithsonian Institution and its staff for assistance in providing many of the subjects for photographs. Thanks are due to Mr W. J. Myatt for contributing valuable Australian material. In the following, S.I. stands for Smithsonian Institution. All photographs are by Lee Boltin, other than:

Pages 2–3—J. Kath, S.I. 6–7—L. Perloff, S.I. 9—Michael Holford, Hamlyn Group Picture Library.

Chapter 1: 13—Agricola, *De Re Metallica,* Basel, 1556. 16—*Hortus Sanitatis,* 1491. 17—John Read, *Through Alchemy to Chemistry,* G. Bell & Sons, Ltd. 20 above left and right— Metropolitan Museum of Art, New York. 20 below—American Numismatic Society, New York. 21 left—Brooklyn Museum. 25—Culver Pictures, Inc. 26—Tom Hollyman, Photo Researchers, Inc. 27 top—L. Perloff, S.I. 29—S.I. 32–3 below—U.S. National Park Service.

Chapter 2: 38—Culver Pictures, Inc. 39—S.I. 42—J. P. Rambosson, *Les Pierres Précieuses,* 1870. 50 left and 51— S.I. 52, 53—L. Perloff, S.I. 55 bottom— Luigi Bombicci, *Corso di Mineralogia,* 1862. 59—Charles Gardner. 62–3 above—J. Kath, S.I. 65—Charles Gardner. 66 above—L. Perloff, S.I. 66 below—W. Ray Scott, Barnaby's Picture Library. 68—British Museum (Natural History).

Chapter 3: 70–1—Boltin, courtesy of Harry Winston. 73 top—A. H. Hind, *An Introduction to a History of Woodcuts,* Dover Publications, Inc., New York, 1963. 73 bottom—Camillus Leonardus, *The Mirror of Stones,* 1520. 74—Pierpont Morgan Library. 75—J. Scott, S.I. 76—Metropolitan Museum of Art, New York. 77—P. H. Groth, *Grundriss der Edelsteinkunde,* 1887. 79—J. Scott,

S.I. 80–1—J. P. Rambosson, *Les Pierres Précieuses,* 1870. 81 right—S.I. 83 top—University Museum, University of Pennsylvania. 84—N. W. Ayer & Son, Inc. 85—Boltin, courtesy of Harry Winston. 86—J. P. Rambosson, *Les Pierres Précieuses,* 1870. 87—South African Information Service. 88—S.I. 89—E. W. Streeter, *Precious Stones and Gems,* 1877. 90—J. P. Rambosson, *Les Pierres Précieuses,* 1870. 91—S.I. 93— J. Scott, S.I. 94, 95—American Museum of Natural History. 96 bottom—J. Scott, S.I. 98—Louvre, Paris. 99—J. Scott, S.I. 100—C. P. Brard, *Minéralogie Appliquée aux Arts,* 1821. 101—J. P. Rambosson, *Les Pierres Précieuses,* 1870. 102—J. Scott, S.I. 103—K. Jensen, S.I. 105 top— British Museum (Natural History). 105 below—Gemmological Institute of America.

Chapter 4: 109—Culver Pictures, Inc. 110, 111—S.I. 113—W. G. Peck, *Introductory Course of Natural Philosophy,* 1873. 116 top—Culver Pictures, Inc. 116 bottom—G. P. Merrill & W. F. Fosbag, *Minerals from Earth and Sky,* 1929. 117 left—Culver Pictures, Inc. 117 right—R. K. Duncan, *The New Knowledge,* 1905. 118—L. Simonin, *Les Pierres,* 1869, 119 top—Paul Seel. 119 bottom, 120–1—S.I. 124—J. Farrell, S.I. 125—S.I. 127 right top and bottom —L. Perloff, S.I. 128—S.I. 129 top— Frank W. Lane, Picture Library. 129 below—South Australian Museum, Frank W. Lane, Picture Library.

Chapter 5: 130–1—William A. Garnett. 133—Bertuch, *Bilderbuch für Kinder.* 135—by permission of the Trustees of the British Museum. 137 top—Van Bucher, Photo Researchers, Inc. 138–9—'Harper's' magazine. 142 left—Ridpath, *Cyclopedia of Universal History,* 1887. 142 right—Australian News & Information Bureau. 143, 146 top—Culver Pictures, Inc. 146 bottom—S.I. 147 top—*Historic Cornish Mining Scenes Underground,* Bradford Barton, Truro, England, 1967. 147 below—F. E. Gibson, Scilly Isles. 148 bottom—Freeport Sulphur Co. 150–1—Chile Exploration Co.

Chapter 6: 155—Radio Times Hulton Picture Library. 160 top—Hamlyn Group Picture Library.

Chapter 7: 166–7—J. Kath, S.I. 169—*Meyer's Grosses Konversations Lexikon,* Leipzig, 1897. 170 top—S.I.

170 bottom—L. Simonin, *Les Pierres,* 1869. 171—'Penny', 1833. 172—Culver Pictures, Inc. 173—A. Wolf, *History of Science, Technology and Philosophy in the 16th and 17th Centuries,* George Allen & Unwin Ltd., London. 174—Bibliothèque Nationale, Paris, courtesy of American Heritage Publishing Co. 175—U.S. Navy. 176—L. Simonin, *Les Pierres,* 1869. 177—U.S. National Park Service. 178—Fritz Henle, Photo Researchers, Inc. 180—Radio Times Hulton Picture Library. 185 bottom—S.I.

Chapter 8: 193—Bettman Archive, Inc. 194—'Journal of Chemical Education'. 195—Swedish Information Service. 196—C. Knight (ed.), *London,* 1842. 197—New York Public Library. 198 left—Glass Container Manufacturing Institute. 198 right—Culver Pictures, Inc. 199—Anaconda Co. 201—Tom Hollyman, Photo Researchers, Inc. 202—Frederick Ayer, Photo Researchers, Inc. 203—UPI. 206 both— Hamlyn Group Picture Library. 207— Bell Telephone Laboratories, Inc. 208—F. Grunzweig, Photo Researchers, Inc. 210 bottom, 211, 212, 213, 214, 215—NASA. 217—Mullard Limited, London.

Chapter 9: 221—Marvin Winter, Photo Researchers, Inc. 224—Jack S. Taylor, Lapidary and Gemstone Supplies, St. Leonards, N.S.W., Australia. 225 top—Paul Seel. 225 centre and bottom—A. E. Goldstein. 228 both —Jack S. Taylor, Lapidary and Gemstone Supplies, St. Leonards, N.S.W., Australia. 229—top Carl Fisher, Photo Researchers, Inc. 229 bottom left and right—A. E. Goldstein. 232, 233—S.I.

Chapter 10: 239—National Gallery of Art, Washington. 240—British Museum (Natural History). 241—Radio Times Hulton Picture Library. 246—British Museum (Natural History).